T0069781

Time Travel
Ten Short Lessons

First published in Great Britain in 2021
by Michael O'Mara Books Limited
9 Lion Yard
Tremadoc Road
London SW4 7NQ

Printed in the United States of America on acid-free paper

2 4 6 8 9 7 5 3 1

Johns Hopkins University Press
2715 North Charles Street
Baltimore, MD 21218-4363
www.press.jhu.edu

Library of Congress Control Number: 2020952491

ISBN 978-1-4214-4240-2 (paperback : acid-free paper)
ISBN 978-1-4214-4241-9 (ebook)

Designed and typeset by Ed Pickford
Illustrations by David Woodroffe

*Special discounts are available for bulk purchases of this book. For more
information, please contact Special Sales at specialsales@jh.edu.*

Johns Hopkins University Press uses environmentally friendly
book materials, including recycled text paper that is composed of
at least 30 percent post-consumer waste, whenever possible.

CONTENTS

INTRODUCTION

Few television shows have had the lasting appeal of *Doctor Who*, first broadcast in 1963 and, despite a hiatus, still going strong over fifty years later. I can still remember the sense of mystery when the first episode went out in the days of black-and-white TV. It was quite different from anything else I'd seen in my young life. And, dramatically, the series was launched during a week that was branded on the memories of anyone old enough to remember it.

At 12.30 p.m. Central Standard Time on 22 November 1963 in Dallas, Texas, US President John F. Kennedy was assassinated. TV broadcasts in many countries came to a standstill. At 5.15 p.m. in the UK the next day, the first episode of *Doctor Who* was broadcast. A combination of power cuts and the shadow of Kennedy's assassination meant that many missed that first episode, so it was rebroadcast the following week. But what made the launch of *Doctor Who* particularly poignant as a series that arrived at such a definitive point in history was that the stand-out feature of the show was time travel.

In part, time travel opened up new vistas. As well as exploring alien planets, the Doctor and his companions could visit historical events. The BBC envisaged this originally as an educational opportunity, though it soon became clear that a visit to the past or future opened up dramatic possibilities and any educational element was given a light touch. Equally, time travel made it possible to explore paradoxes when, for example, it became feasible to go back in time before a disaster and change things so that it never happened. In practice this has rarely featured in the programme, but time travel is a tempting snare for the imagination.

These paradoxical possibilities would feature more strongly in some Hollywood ventures into time travel, from *Back to the Future* to *Looper* and beyond – and, of course, in literary classics such as Ward Moore's *Bring the Jubilee* and Douglas Adams' *Life, the Universe and Everything*. But, generally speaking, the audience for science fiction has taken time machines to be a fun convention that allowed for an illusory ability to be played out, because the assumption for many was that time travel would never be possible in reality.

Although science fiction, unlike fantasy, tries to stick to what is physically possible, it has always featured a handful of special conventions where something that is believed to be infeasible is allowed to happen. The rules allow these unlikely concepts to be introduced, after which the storyline has to follow what is practical within the known world. An early example of this was in the H. G. Wells novel *The First Men in the Moon*. Wells introduced an imaginary substance, cavorite, which blocked gravity. But once he had brought this impossible invention into the storyline, the consequences moved forwards logically.

Such frequently used cheats have included faster-than-light travel, hyperspace, force fields or 'shields', tractor beams and, of course, time travel. And in most of these cases, it is hard to see how the technology or concept could ever be made real. But time travel is the exception that proves the rule. Because, remarkably, Einstein's theories of relativity have made it clear that making a time machine is more a matter of engineering challenges than physical impossibilities. Time travel is real and is happening right now.

Travelling through time has a seductive appeal for practically everyone. Who wouldn't be fascinated by a visit to the past? Admittedly, it would be a problem if you worked in the gambling industry: there would be no future in lotteries or betting if anyone could pop back and place a bet on a certain outcome. But for the rest of us, it's a beguiling prospect. History and archaeology have their limitations. It would be amazing to be able to watch or take part in great events from history and see what really happened, or to see living dinosaurs striding across the Earth, no doubt turning palaeontology on its head. For that matter, on a personal level it would be transformative to have the chance to see dead relatives again, to revisit things you never got the chance to say. And then there's the future – whether that be dark or bright. The reality of things to come would be science fiction come alive, a heady plunge into a speculative world.

So let's take a step beyond the Doctor, Doc Brown and all those time-travelling fictional characters and take a real-world trip in the fourth dimension.

01 TIME TRAVEL IS MORE THAN FICTION

'Clearly any real body must have extension in four directions: it must have length, breadth, thickness, and – duration.'

H. G. WELLS (1895)

It's no surprise that time travel crops up regularly in fiction, but it is quite shocking to learn that it can be done for real. In one sense, we all travel in time. Our conscious moment of the present seems to tick forwards, gliding from hour to hour and day to day. And, thanks to memory, we can slip back into the past, revisiting another time – even if memory is now recognized as a mental construct that inaccurately recreates what once was. Yet this picture of time travel seems like cheating. It's not what we hope for. It's a bit like saying we are physical travellers sitting in a chair at home because the planet Earth is constantly in motion around the sun. Thankfully, though, it's only the beginning.

This is because true time travel – the ability to move to a time other than the present based on science and technology – is real. There is nothing in the laws of physics that prevents time travel, and the science behind it has been experimentally proved many times over. We can, with H. G. Wells' time traveller, take on a journey in a fourth direction that is different from the spatial three. A trip through the dimension of time.

Traversing the dimensions

Every experience you have involves movement through space. It may be your body moving, or something more subtle: you couldn't see, for example, without photons of light moving from the source to your eyes, breathe without air molecules moving into your lungs or even think without the movement of electrical impulses in your brain. And all movements involve some elapse of time: nothing can move instantaneously. That's a reality that was challenged in the fifth century BC by an ancient Greek philosopher called Zeno. According to Zeno and the Eleatic school, change and movement were nothing more than illusion. Zeno illustrated this viewpoint with a number of paradoxes.

In one of Zeno's most famous examples, an arrow is flying through space towards its destination. Imagine taking a look at that arrow at a single

❝ Time forks perpetually toward innumerable futures. ❞

JORGE LUIS BORGES
(1958)

moment in time, comparing it with another arrow that isn't moving but is simply hanging in mid-air (let's not worry about how). What's the difference? In that instant in time, neither arrow is moving. So, can movement be a real thing if we can't distinguish it from stillness in any particular moment?

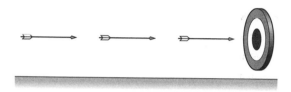

Zeno's paradox arguably falls apart because, outside the philosopher's world, there are no fixed moments in time. We can't stop time or reverse it. Time ticks on regardless. However, the paradox illustrates a fascination with the dimensions of time and space – the essential components of travel.

Although we can move freely in the three spatial dimensions (subject to physical objects and gravity getting in the way), we can only dream of moving at will through a fourth dimension – time – choosing to visit the future or past rather than inhabiting the familiar present. All our experiences are limited to the now. But that hasn't prevented writers from envisaging the ability to break out of the present and treat time as if it were indeed a fourth dimension.

HUBERT GEORGE WELLS (1866–1946)

Although never formally having studied physics, Wells was clearly fascinated by the nature of time, and in 1888 the twenty-two-year-old published a short story in the student newspaper, *The Science Schools*, entitled *The Chronic Argonauts*, which featured a machine that made time travel possible. Gaining an external zoology degree from the University of London in 1890, Wells became a prolific journalist over the next few years, writing both articles and short stories. He initially hoped to sell a polished version of *The Chronic Argonauts*, but instead developed in 1895 a more sophisticated novella with a similar device, *The Time Machine*, first published as a serial in *The New Review*. Other prominent science-fiction novels followed, including *The Invisible Man*, *The War of the Worlds*, *The First Men in the Moon* and the near-unreadable but influential *The Shape of Things to Come*. Wells would also have success with everything from romance to non-fiction. He revisited time travel in 1899 with *The Sleeper Awakes*, using a more traditional approach of putting his hero to sleep for 203 years.

The earliest accounts of such travel were matters of magic, fantastical voyages that had no more scientific justification than had mythological trips to the sun or moon. A traveller might be transported to a different time by an angel, committed to decades of magical sleep with the prick of Sleeping Beauty's sharp spindle, or conveyed to another era by a bump on the head, as was the case with the protagonist in Mark Twain's whimsical *A Connecticut Yankee in King Arthur's Court*. But by the time Twain wrote his time-travel novel in 1889, the scientific future was looming large in creative minds, and just six years later, H. G. Wells' time traveller would be one of the first to deploy technology (however vague in its mechanism) in a journey through the fourth dimension.

> **The distinction between past, present and future is only a stubbornly persistent illusion.**
>
> ALBERT EINSTEIN (1955)

The time-travelling writer's toolbox

Read histories of science fiction and you will find a range of dates for the beginning of the genre. Some reach back to Mary Shelley's 1818 *Frankenstein*, others to the 1638 *The Man in the Moone*, published after the death of its author, bishop of Hereford Francis Godwin. Here, the protagonist, Domingo Gonsales, makes the journey to the moon towed by a special breed of migrating birds. At the most extreme, we are directed to Lucian of Samosata, a Greek-speaking Roman living in Syria. Lucian's *True History*, written in the second century AD, was a parody of the *Odyssey*, a fanciful story that sent its protagonists to the moon, lifted up by a whirlwind. But many would argue that true science fiction began with the two Victorian giants of the field, Jules Verne in France and H. G. Wells in England.

> **Time is an illusion, lunchtime doubly so.**
>
> DOUGLAS ADAMS (1978)

Verne wrote speculative fiction that was driven primarily by engineering, while Wells was more a writer fired by the imagination. Verne mocked this distinction, commenting of his significantly younger competitor's writing: 'I make use of physics. He invents. I go to the moon in a cannonball, discharged from a cannon. Here there is no invention. He goes to Mars in an airship, which he constructs of a metal which does not obey the law of gravitation.' Leaving aside Verne's confusion of two Wells novels – *The War of the Worlds* involving Mars and *The First Men in the Moon* where the fictional anti-gravity metal cavorite was used – what Verne underlined here is that there was already more than one type of science fiction, with some stories more speculative than others.

The distinction between Verne's and Wells' visions was, however, not as clear cut as Verne suggested. Verne might have stuck to the capabilities of engineering, but in practice the acceleration from his space cannon would have squashed his astronauts to paste. Wells certainly did make use of imaginary concepts, for example his mystery metal cavorite, but once he had introduced them, he employed physics as we know it. In doing this, Wells prefigured what would be allowed later in 'hard' science fiction – stories that stick as much as possible to the laws of physics: the employment of a small number of tropes without knowing how they might be made possible. Perhaps the most common is faster-than-light space travel to enable journeys to the stars. But equally important is time travel – something that, as we have seen, Wells really began – simply to revel in its possibilities.

There is a wide range of time-travel mechanisms and outcomes that fiction would go on to explore. As we have seen, some stories use an extended sleep to get into the future, while magical means have continued in books such as Audrey Niffenegger's *The Time Traveller's Wife*. But Wells changed time travel for good from a vague, ethereal excursion to something that would be carried out using science and technology. The word that was most significant in the title of his novel was 'machine'.

Wells used *The Time Machine* primarily to explore the way that the social divisions of the day that he condemned might grow worse. But the stories that followed on from Wells' work would introduce new possibilities derived from the technology of time. Fictional time travel could

FIVE GREAT TIME-TRAVEL STORIES

1895 | *The Time Machine* | H. G. Wells
Wells' time traveller visits AD 802,701 where society
is split between the delicate Eloi and the brutish
Morlocks, then 30 million years ahead to a dying Earth.

1952 | *A Sound of Thunder* | Ray Bradbury
In this short story, hunters travel into the far past
to kill dinosaurs just before they would die anyway.
As a result of a crushed butterfly, the future is
irrevocably changed.

1955 | *The End of Eternity* | Isaac Asimov
Typical of a class of novel featuring 'time police' whose
job is to keep the timeline from being distorted:
politics, personalities and the manipulation of time
result in the organization ceasing to have ever existed.

1959 | *All You Zombies* | Robert Heinlein
The most twisted of bootstrap paradoxes where the
same person turns out to be every major character in
the story, including his own parents.

2012 | *Looper* | Rian Johnson
Assassins are sent back into the past, finally killing
their older self. Features altered timelines and one of
the most scientific time-travel technologies in fiction.

involve, for example, the intertwining of two or more timelines in a time slip. There can be paradoxes arising where a temporal loop is formed, breaking the familiar relationship of cause and effect. There are the mind-boggling 'bootstrap' paradoxes in which time travel enables something to come from nothing. Then there is the danger of the so-called butterfly effect, where a small change in the past can result in major revisions of the future. And always there is the sense of wonder and clash of culture when people from two different times meet. Who could not resist the enjoyment of turning up, say, in the Victorian era with twenty-first-century technology?

When Wells had the traveller in *The Time Machine* point out that time is the fourth physical dimension, he was making a point that would become far more significant ten years later, when Albert Einstein burst onto the scene with his special theory of relativity. It is relativity that would provide the foundation for all the science of time travel that involved more than simply sleeping for a number of years to reach the future. It's easy to think of Einstein as the originator of relativity: his name is so strongly associated with it. But in reality, relativity

is a much older concept. Galileo Galilei established the basics of relativity nearly three hundred years before Einstein.

Everything is relative

Relativity emerges from an understanding of the nature of movement. Are you moving as you read this? Your answer would probably depend on where you are located. If you are sitting in a chair at home, no. If travelling on a train or plane, yes. But in saying this, you are being misled by the presence of a nearby extremely large object: the Earth. What you really mean is that you are moving (or not) *relative to* the Earth. But bear in mind that, with the Earth, you are hurtling around its orbit of the sun at around 100,000 kilometres (62,000 miles) per hour. And, along with the sun, the Earth is travelling around the Milky Way

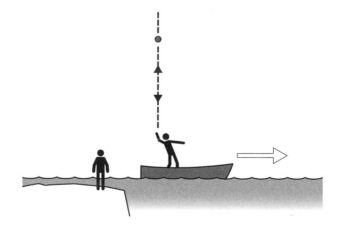

at an even brisker 800,000 kilometres (500,000 miles) per hour. Whether or not you are moving depends entirely on how you make the measurement – what you arbitrarily consider to be fixed in place.

And this is what Galileo realized. There is no master fixed grid in the universe against which all movement can be measured. We have to pin down what physicists call the 'frame of reference' against which we measure that movement. Galileo showed that an object thrown upwards in a steadily moving boat would still fall straight back down: in fact, if the boat had no windows it would be impossible to determine it was moving. Relative to the boat, there would be no motion. Isaac Newton would make use of some aspects of Galilean relativity in his laws of motion nearly a hundred years later but, despite this, it was not until the end of the nineteenth century that the implications of the lack of absolutes in time and space were fully realized, coming to fruition in Albert Einstein's special theory of relativity.

> **Henceforth, space by itself, and time by itself, are doomed to fade away into mere shadows.**
>
> HERMANN MINKOWSKI (1979)

Einstein's revolutionary theory added in one extra factor to Galilean relativity. Unlike his predecessors, Einstein knew that there was something special about light. One of Einstein's heroes, the Victorian Scottish physicist

James Clerk Maxwell, had identified light as an interplay between electricity and magnetism – an effect that could only happen at one specific speed in any particular environment. The speed of light cannot be altered by movement. Unlike everything else, light's velocity is not relative. Travel towards or away from a beam of light and it still blasts towards you at the same 300,000 kilometres (186,000 miles) per second in a vacuum (it's a bit less in air). And with that small but significant addition to relativity, Einstein found that time and space had become intertwined. Movement through space has an influence on the flow of time. Here was the key to manipulating time's apparently constant progress.

Although Einstein's concern in developing his theory was to forge an understanding of the nature of reality rather than creating anything practical, he had opened up the opportunity to travel in time through its relationship with space. A few years later, Einstein would return to relativity in developing his general theory, which took in the twin aspects of acceleration and gravity. This would provide the solution to a puzzle that had been around since the time of Newton: how gravity allowed one object to influence another at a distance. This was

> We think we know what time is because we can measure it, but no sooner do we reflect upon it than that illusion goes.
>
> ROBERT MACIVER (1921)

something that Newton's critics had found so strange that they referred to his theory as 'occult'. But the general theory of relativity also provided the final piece in the puzzle of making time travel a reality. And at the same time, Einstein's work gave new impetus to the tellers of tales who came after H. G. Wells. Science fiction had a new reality to explore.

The fiction of time travel is both fun and informative. Science fiction writers were able to explore the implications and paradoxes of time travel long before physicists were capable of experimentally testing the theoretical possibilities that had been raised by Einstein's theories of relativity. But the story of time travel is no longer primarily one of fiction. Once Einstein opened the door to real travel through time, there was no going back. The fourth dimension was freed up. Time travel was physically possible – and has been demonstrated many times since Einstein came up with his theory, sat at his desk in the Swiss Patent Office in Bern.

But it is too soon to join him there: we are getting ahead of ourselves. Time is a slippery customer. It is surely premature to begin thinking about the possibilities of time travel without first pinning down the nature of time itself.

02 HOW TO UNDERSTAND TIME

'Time goes, you say? Ah no! Alas, time stays and we go.'

HENRY AUSTIN DOBSON (1877)

'Time' is a word we make use of all the … time. According to the Oxford Dictionaries website, time is the fifty-fifth most common word in use in written English today – which sounds quite impressive, but is significantly more so when you realize this makes it the most commonly used noun.

There's no doubt that time is something of an obsession. Our smartphones, computers, watches and more all keep us up to the minute on its passing. All this technology might suggest that clock-watching is a modern preoccupation, but mechanical clocks have been around since the end of the thirteenth century, and before that, water clocks and sundials were used – or simply the passage of the sun through the sky alerted people to the progress of the hours. Perhaps the most impressive acknowledgement of how long time has been an obsession comes from the fourth century bishop St Augustine of Hippo. He remarked:

> What is time? Who can explain this easily and briefly? Who can comprehend this even in thought so as to articulate the answer in words? Yet what do we speak of, in our familiar everyday conversation, more than of time? We surely know what we mean when we speak of it. We also know what is meant when we hear someone else talking about it. What, then, is time? Provided that no one asks me, I know. If I want to explain it to an inquirer, I do not know.

Hippo was a Roman city in North Africa, now Annaba in Algeria. Augustine's words were written around the year

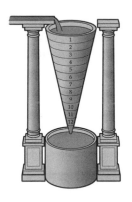

400. We might think of the attitude to time back then being much more relaxed, but it's fascinating that Augustine considers time a frequent topic of conversation. And his analysis of the nature of time is extremely clever. We all know what we mean by time – but it is practically impossible to explain to someone else what time is.

AUGUSTINE OF HIPPO (354–430) ON TIME

Born in what is now the seaport of Annaba in Algeria, Augustine wrote about time in *Confessions*, mostly a work on theology, but one in which he also explored his own past and the nature of reality. In *Confessions*, Augustine noted that time 'tends to non-existence', referring to the way that we only truly experience the now, rather than an expanse of time, making it feel more like a direction than a true dimension (this is something that is echoed in Eddington's concept of the 'arrow of time' described on page 28).

Apart from his wry observation concerning the difficulty of pinning down what time is, Augustine also pointed out that, if the future or the past actually existed separately from the present – in effect as a destination that could be reached – then it wasn't clear exactly where these 'places' existed, as if we were to visit them (my words, not his), they would be the present, not the past or the future. Our very existence defines the present.

What is time?

In trying to pin down the nature of time, we are faced with something similar to the problems facing biologists who attempt to define what life is. They can come up with a series of properties, such as nutrition or reproduction, which are common to most or all living things, but it is extremely difficult to say what life *is*. As Augustine complained so eloquently, the same goes for time.

There is no particular help in moving from an ancient bishop to a modern physicist. Although you'll find plenty of popular science titles such as Stephen Hawking's *A Brief History of Time* or Lee Smolin's *Time Reborn* that seem to promise they will tell us about time itself, most skirt around what time is and focus on the relationship between time and space. Even Carlo Rovelli's *The Order of Time*, which does indeed attempt to reveal the nature of time, only manages to do so in an indirect, poetic fashion and often contradicts itself, frequently telling us that time does not exist (we'll come back to this idea), while elsewhere saying, 'Time and space are real phenomena.'

> **We inhabit time as fish live in water. Our being is being in time.**
>
> CARLO ROVELLI (2018)

While not being able to pin down time too closely, we can identify three roles that it fulfils. We'll start with Wells' description of time as a fourth dimension. Imagine the three dimensions of space with a fourth dimension at right angles to all of the rest. That, admittedly, is a bit of a strain on the imagination. But we can simplify things by ignoring one of the spatial dimensions – or even better, two of them, so that we just have two directions, one in space and the other in time. We might make space horizontal and time vertical.

We can now plot the position of an object in space and time on a simple chart, known as a Minkowski diagram. In this picture of dimensions, a physical object that isn't moving (with respect to an arbitrary frame of reference – the Earth, say) would simply be represented on our chart by a line going straight up the time dimension. We already have a mechanism for defining a position on the face of the Earth – latitude and longitude – a pair of numbers that enable us to pinpoint a location. In the first of the three roles of time, it provides us with the equivalent locator on the time dimension. It's what's known as a coordinate system – a way of identifying exactly where you are, in this case your location in time.

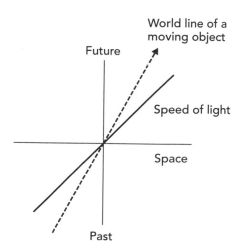

There's something else you can do once you have a way to pin down a location – and this provides us with the second role of time. If you have two locations in space, you can work out the distance between them. Similarly, given two points on the time dimension, you can calculate a temporal duration: how long something takes, or for that matter, how long you have to wait for an event in the future to arrive – your next birthday, for example.

As I write this, according to my computer's clock it is 11.08 on the morning of 27 April AD 2020. Making this statement effectively brings in both the first and second roles

> ❙ **Time is what keeps everything from happening at once.** ❚
>
> RAYMOND KING
> CUMMINGS (1929)

of time. It gives my position on the time axis, but the label given to that coordinate is entirely arbitrary. The date is also Ramadan 4, 1441 AH, for example, or 3 Iyar, 5780 in the Hebrew calendar. And though the time where I am in England is 11.08, it could be anywhere from 00.08 to 06.08 in the USA, from 17.08 to 21.08 in Australia, or even 16.53 in Nepal, where the time zone is not an exact hour's difference from most of the rest of the world.

None of these times or dates is 'right' or 'wrong'. Each reflects the reality that our measurement of the coordinate of time is not an absolute value, but a relative position on the time dimension. We decide on a key point in history and then measure the time *now* by using the second role of time – the duration of time that has passed since that point. In this view, time is what clocks measure. On our two-dimensional plot of space and time, this is a measurement of distance along the time axis from that arbitrary fixed point where the axes cross.

Computers do something similar, often storing time as a number of seconds since, say, 1 January 1900. If you are old enough to remember the Millennium Bug scare in 1999, there was a concern that some computer programs had not taken into account that they would still be used in 2000, so only made use of a limited space to store the date, which it was thought might run out and crash the computers. In practice it rarely proved a problem, but it's an example of the way that the arbitrary nature of coordinates in time have the potential to intrude into the real world.

Does time flow?

The third role of time comes across as a reflection of something that appears to progress and flow. We poetically refer to the river of time or time's arrow. The past is behind us, the future in front. We have a sense of constant motion through time, even though in reality all we ever experience is the moment that is 'now'. In effect, we experience the flow of time as we do a car journey looking out of the back window. We don't really experience motion itself. With respect to the car, we aren't moving. But the outside world flows past, providing a constant addition of new moments of time to our memory of the past.

It is this motion through time that some physicists dismiss as a purely subjective concept. Apart from anything else, such a movement offends the physicist's sense that to be meaningful (to physics), something should be capable of being measured. We seem to move through time at a rate of a second per second – but even a moment's thought suggests that this is a meaningless concept. For some physicists, a clearer image is what's known as the block universe. Let's think back to our representation of the existence of an object – you, for instance – as a trace on a chart with axes representing space and time. If there is indeed a flow of time, we would see your existence as a point that moves on that chart, steadily travelling up the time axis. But in the block universe there is no motion. The Earth and moon, for example, would be continuous objects, stretched out through time.

When physicists say (as some do) 'there is no such thing as time', it's the idea that time flows to which they are referring. Leave them without food long enough, and they will be just as supportive as any normal person of the idea that the

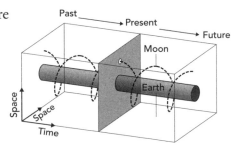

time to eat exists, and it's now. But they would deny any kind of movement from the point in time when they were satisfied to the point where they began to feel hungry.

Underlying this claim is the observation that most physical processes don't really care about the flow of time.

> It's no use going back to yesterday, because I was a different person then.
>
> LEWIS CARROLL (1865)

Think, for example, of a collision between two spaceships. We can imagine a video of the ships heading towards each other, colliding and bouncing off, then heading back in the directions from which they came. If the spaceships were able to just bounce off each other without distorting in any way, the result would be, in physicists' terms, reversible. You could run the video backwards and there would be no distinction. It would look exactly the same.

I had to come up with a fairly artificial example to illustrate this, because real life is rarely like physics examples, which usually have to be extremely simplified. (An infamous physics simplification is 'Let's assume that the cow is a sphere.') It's unlikely in the real world that there would be no damage, so the 'after collision' ships would look different to the 'before collision' ones. Some heat would be generated by the collision, which means that the two spacecraft would bounce away from each other at a slightly lower speed than that at which they arrived, as the rest of their kinetic energy of motion would have gone to produce that heat. And, in most of the real occurrences around us on Earth, there would be friction and gravitational effects that might mess up the symmetry of the experiment (this is why I put the example in space).

So, despite the claims of those physicists, the real world does seem to have a direction of time along which we can meaningfully think of a flow. And underlying this concept is the innocuous-sounding second law of thermodynamics.

The second law and time's arrow

Note that in my best possible example of a collision in space, there would still be heat generated. The second law broadly states that, statistically speaking, things get more disordered with time. It's why, for example, it's much easier to break a glass than it is to unbreak it. And after our experiment, the heat generated would mean atoms were moving around more, increasing the disorder.

The result is what is usually referred to as the 'arrow of time', following English physicist Arthur Eddington's suggestion – the second law of thermodynamics indicates a particular direction in the time dimension that provides the natural order of things. In practice, even if simplistic physical models are reversible, actual time isn't. So, though

THE LAWS OF THERMODYNAMICS

The laws of thermodynamics are among the central pillars of physics. Although originally developed to describe the flow of heat in steam engines and other heat-based technology, they are far more fundamental, usually having more than one way to be considered.

Law	Formulation	Alternative formulation
Zeroth law	Two objects are in equilibrium if heat can flow between them but doesn't.	If two systems are in equilibrium with a third they are in equilibrium with each other.
First law	Energy is conserved in a closed system.	The energy in a system changes to match the work done and the heat exchanged.
Second law	In a closed system heat flows from hotter to cooler parts.	In a closed system entropy will stay the same or increase.
Third law	A body cannot reach absolute zero.	The entropy of a system approaches zero as the temperature approaches absolute zero.

> ❝ I shall use the phrase "time's arrow" to express this one-way property of time which has no analogue in space. ❞
>
> ARTHUR EDDINGTON
> (1927)

the block universe may be the best way to represent the relationship between space and time, we have to acknowledge that the block has at the very least an arrow embedded in it saying 'This way to the future' – although the concept of time flowing may be subjective, it's as good a way to describe what's happening as any.

That the flow is subjective is underlined by the variability in the way that time seems to pass by. Most of us are aware that when we were children, time seemed to flow extremely slowly – sometimes at an agonisingly slow pace. It was so easy to get bored. As we get older, though, subjective time passes more quickly. Equally, what we are doing during a period of time will strongly influence how that time seems to pass. If you find a movie dull, for example, it can seem to last for ever, but if the story truly engages you, a couple of hours can disappear surprisingly quickly.

Albert Einstein once claimed to have undertaken an experiment on the passage of time, helped by silent film star Paulette Goddard, who he had met through mutual friend Charlie Chaplin. Einstein summed up the 'experiment' in the abstract of a paper: 'When a man sits with a pretty girl for an hour, it seems like a minute. But let him sit on a hot stove for a minute and it's longer than any hour. That's relativity.'

The paper in question did not exist. Einstein never performed an experiment in his working life (though he may have spent some time sitting with Paulette Goddard). It's sometimes missed that the initials of the publication in which he claimed that the paper appeared spell out an appropriate word: it was the *Journal of Exothermic Science and Technology*. And Einstein's throwaway line about relativity was humour, rather than a scientific assessment, as his theories of relativity would provide a mechanism to allow objective, rather than subjective, time travel. But there is no doubt that the way we experience the flow of time is a very personal one.

Apart from anything, that apparent flow is not continuous. Our conscious appreciation of the flow of time can make sudden leaps, rather than progressing smoothly from place to place on the time dimension. Every night (sleeplessness apart) each of us makes one or more leaps from point to point in time without experiencing any of the time in between. And that leads us to the most immediately available

> **We don't measure time with seconds, like our clocks, but by our experiences. For us, time can slow down or time can fly.**
>
> AINISSA RAMIREZ (2020)

– if distinctly risky – form of time travel. Stopping your consciousness for an appropriate period of time to be able to jump into the future.

03 HOW TO BE A CORPSICLE

'I wish it were possible, from this instance, to invent a method of embalming drowned persons in such a manner that they may be recalled to life at any period however distant … I should prefer to any ordinary death the being immersed in a cask of Madeira.'

BENJAMIN FRANKLIN (1773)

In an industrial estate just off the Edsel Ford Freeway in Clinton Township, Michigan, is a modest-looking building, identified by its signage as the Cryonics Institute. Inside are a series of 3-metre (10 foot) high cylinders known as cryostats, each a giant vacuum flask, holding up to six human bodies in liquid nitrogen at –196 °C (–320 °F). Cryogenic facilities like this one reflect an attempt at the most basic form of time travel: we all travel into the future every moment – the trick of effortless forwards time travel is to get there quicker from a subjective perspective.

As mentioned in the previous chapter, we all do this when we sleep. Most nights we will jump forwards several hours with only conscious awareness of dreams. And individuals in comas have jumped forwards several months this way. When waking after being unconscious, we have indeed travelled in time – but we will also have aged. To make this a worthwhile mechanism, a time traveller wants to avoid the gradual decay of the flesh.

Those who have their bodies placed in cryogenic (the general term for low-temperature work) storage as soon as possible after death take the gamble that at some time in the future scientists will be capable of restoring them to life, with their memories and consciousness intact. They must also hope that the company storing them will last long enough for such a technology to be developed. Many opt for whole body storage, but others go for the cheaper option (because it takes up less room) of just storing the head, on the assumption that artificial bodies will eventually be available.

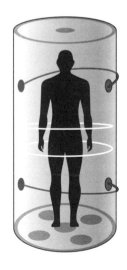

When Benjamin Franklin made his observation about wanting to be immersed in a cask of Madeira it was after claiming that he had seen three flies restored to life after being dried out in the sun, despite having spent many months in a bottle of Madeira. Franklin's flies seem unlikely in reality, but we do know that small creatures known as tardigrades can survive being dried out and apparently come back to life.

Also known as water bears, tardigrades are tiny insect-like animals around 2 millimetres (0.08 inches) in length. They are extremely hardy creatures, which can survive in a wide range of conditions, even having survived several days exposed to the vacuum of space. When in a dehydrated form, they can be revived years later. Tardigrades certainly demonstrate that, with the right protection, cells can survive without the normal conditions of life, in their case due to a special internal composition. However, tardigrades are not dead in this state – they are, in effect, in an extreme version of sleep where they do not exhibit many of the normal signs of life.

Sleeping into the future

The idea of using sleeping as a vehicle for reaching the future was nothing new when American essayist Washington Irving (then living in Birmingham in the English midlands) published his short story *Rip Van Winkle* in 1819. A number of myths and legends exist in which individuals are tempted into strange lands where time passes differently to our own, or groups undergo a magical sleep, locked away in a secret location until they are required. A good example of this is the legend of Alderley Edge in Cheshire, where a group of warriors and their horses, sometimes associated with King Arthur, are said to be sleeping until England is at its time of greatest need.

Such travel into the future by sleep is also implied in the familiar fairy story of *Sleeping Beauty*, in which the central character (depending on versions) is said to have been asleep for up to a hundred years – though the traditional story, which seems to date back at least as far as the fourteenth century, pays little attention to the fact that the protagonist has awoken in the future. Arguably, in medieval times, change was so slow that there would have been little difference for the sleeping princess to experience, other than the loss of people she knew. But for the sleeper, dreams apart, there would be an instant leap as her conscious self is restored to the waking world.

> **What a wonderfully complex thing! This simple seeming unity – the self! Who can trace its reintegration as morning after morning we awaken ...**
>
> H. G. WELLS (1910)

Science fiction would also make use of this approach for reaching the future, starting with H. G. Wells' *When the Sleeper Wakes*. By the time this novel was written in

FIVE TIME-TRAVEL SLEEPERS IN FICTION

14th century?	*Sleeping Beauty*	Traditional	A princess is put into a sleep by a poisoned spindle until kissed by a prince.
1819	*Rip Van Winkle*	Washington Irving	The eponymous Winkle falls asleep for twenty years after drinking magical liquor.
1899	*When the Sleeper Wakes*	H. G. Wells	The central character overdoses on insomnia drugs and is in a coma for 203 years.
1931	*The Jameson Satellite*	Neil R. Jones	A professor's corpse is frozen until he can be revived after millions of years.
1998	*The First Immortal*	James L. Halperin	A man who dies in 1988 is cryonically preserved and is revived and modified to live indefinitely.

1899, then revised in 1910 as *The Sleeper Awakes*, the possibilities for change in both technology and society that Wells had highlighted in his earlier book *The Time Machine* were far more apparent than had been the case before the Industrial Revolution. But Wells' hero, Graham, is still the subject of an accidental sleep into the future, entering a coma in 1897 and not reviving until 2100. It was only later that it was thought possible that someone might be intentionally put into some kind of suspended animation to pass through the years without ageing.

In science fiction, this approach has often been used as a way to undertake otherwise impractically long journeys in space. We see, for example, three crew members cooled to put them in a form of suspended animation on the way to Jupiter onboard the *Discovery One* in the classic 1968 movie, *2001: A Space Odyssey*. But as yet, at least, in the real world, long-range human space travel is not occurring. Instead, the development of cryonics has been based on the idea of preservation of the body at the point of death in the hope of a future society being able both to revive the individual and to cure whatever caused the death.

Frozen until fixed

This idea also began to turn up in twentieth-century fiction, for example in the 1931 short story *The Jameson Satellite* by Neil R. Jones, which featured a professor whose corpse was frozen in orbit for millions of years before being revived. This story seems to have inspired a 1962 non-fiction book, *The Prospect of Immortality*, written by American physics teacher Robert Ettinger. From his mid-forties when he wrote the book, through to his death (and inevitable cryonic preservation) in 2011, Ettinger was an advocate for the idea that a body could be sufficiently well preserved for it to be revived when technology was advanced enough to do so – by which time he believed it would be possible to keep people alive for ever (hence the book's title).

Cryonics has always straddled the boundary between science and science fiction. Its origins were amateurish, reminiscent of the garage enthusiast origins of personal computing. One of the first groups to perform freezing, the Cryonics Society of California, which was the personal domain of a one-time TV repairman, had trouble keeping solvent. The organization resorted to putting several bodies in the same container, at one point losing nine of its clients when two of the systems broke down. In the early days there was also little understanding of the mechanisms that would be required to preserve tissue at low temperatures without damage.

> **Although no one can quantify the probability of cryonics working, I can estimate that it is at least 90 per cent …**
>
> ARTHUR C. CLARKE (1988)

Modern cryonics organizations, such as the Cryonics Institute and Alcor in Arizona, have moved the technology on a generation or two from those early days. Although the low-temperature storage environment is still much the same, the preparation of the body is much more likely to keep cells intact, making it at least in principle possible to revive a preserved individual, even if there remain doubts that the electrochemical processes that make you the individual you are could possibly survive this treatment. Cryonic preservation involves the replacement of blood with a 'cyroprotectant', which prevents the damaging formation of ice crystals that destroy the integrity of the cells in straightforward freezing.

The process now used is known as vitrification. Proponents of cryonic preservation make a distinction between vitrification and freezing because the approach reduces the chances of crystals forming. (Vitreous means glass-like, and glass is an 'amorphous' solid that does not contain crystals.) The preservation material does still solidify at the ultra-low temperatures that are used, but in a less damaging way than a water-based medium would. If vitrification is carried out correctly, the cells of the body should become a glass-like solid structure.

This process is used for human egg and embryo preservation, and would be attractive in organ donation, as at the moment a donor organ can only be preserved relatively briefly, but a vitrified organ could be stored for months until required. Vitrification has been carried out successfully with a rabbit kidney, but as yet it is not viable for human organs. However, a human brain has a far more complex structure than any other organ in the body, making it questionable how likely it is that a revival would be possible.

Is there anybody in there?

How likely it is that anyone really could be revived after cryonic storage is disputed. Neuroscientist Michael Hendricks has argued that the technology simply cannot deliver what is promised – and his viewpoint is echoed by many mainstream scientists. Hendricks points out that your consciousness is not just a matter of your brain's physical structure, but also the electrochemical linkages in a state of continuous flux, something that cannot reasonably be preserved in dead brain cells.

Representatives of companies such as Alcor protest that they are not storing dead people. This is because the more advanced cryonic facilities try to keep the body alive using life support until the temperature is reduced, even though the individual is considered medically beyond retrieval. The cryonics organizations claim to intercept the dying process before it becomes irreversible. However, this does feel like sophistry. The fact remains that the vitrified body (or head) is not alive – they may have been vitrified while technically alive, but the vitrified remains have none of the characteristics of life. The 'Cryonics Myths' section of Alcor's website says: 'Cryonics is not a belief that the dead can be revived. Cryonics is a belief that no one is really dead until the information content of the brain is lost, and that low temperatures can prevent this loss.' While the first part of that belief has a potential scientific basis, the second part – that low temperatures can prevent the loss

> ❚ **The individual who freezes himself or herself to come back in the future makes the assumption he will be a contributor to society and that they would want him.** ❚
>
> JOHN BAUST
> UNESCO Professor of
> Biological Sciences (2002)

of information content from the brain – has no evidence to back it up. It's a hope, rather than a scientific fact.

The other issue that has to be faced as far as using this approach to time travel goes is that it puts considerable faith in the future, both in terms of finance and of ethics. It is not just a matter of having the technology to bring a cryonically stored body back to life and consciousness. The business of storing the body will have to continue to be economically viable for what is likely to be hundreds of years into the future. And even if that is the case, the people of the future have to have a motive to restore the stored body. In Wells' *The Sleeper Awakes*, the Sleeper is a one-off, a mythical creature seen as a kind of saviour of humanity.

But we have to ask how much a future society would have interest, beyond a one-off freak show, in restoring individuals from the past. Imagine we had a means of restoring someone from Tudor times. It would be fascinating, certainly, and we would want to do it. But if hundreds or thousands of people from several hundred years earlier were available, would they be able to integrate into modern society? Would they have skills that would make them desirable beyond the novelty factor? There is not a cut-and-dried answer. Even more so, if, as Robert Ettinger envisaged, the technology of the future meant that people would live forever, would the inhabitants of the future want to add these relics to an already crowded world?

Alcor suggests that this viewpoint, 'suggesting that humans have no intrinsic value, but only have value based on whether they "contribute to society" or whether others "want" them', is ethically questionable. Perhaps it is. Yet if we are to venture into ethics, and should this argument be accepted, it would be a lot simpler to rescue some of the millions of children who die each year in developing countries. We only have to look at America or the EU's attempts to stop migrants crossing their borders. It is perhaps a stretch to think that things will be any different when we are considering refugees attempting to cross the border of time into the future.

For the moment, cryonic preservation remains very much a minority interest. Facilities typically store fewer than a hundred individuals, and even Alcor, the largest in the world, has well under two hundred stored clients in a mix of full body and the cheaper head approach, though a few thousand are apparently signed up for future storage. It seems quite a low-probability route to travel into the future.

Uploading to the future

If the cryonic processes sound too gruesome or unlikely to succeed, others hope they will soon be able to travel into the future by uploading – storing a copy of their brain structure on a computer. Although this is not yet possible, computer technology continues to advance at an exponential rate. In the future, this could make it possible for a personality to continue in an electronic medium, even though the biological original would still die. Once uploaded, time travel would be easy – simply a matter of suspending consciousness for whatever period was required.

However, neuroscientist Michael Hendricks is as doubtful of the idea of uploading as he is of cryonics, if not more so. He points out that his main study is of *Caenorhabditis elegans*, a small worm whose whole neuronal network is the best-studied such structure in the field. Yet even a complete simulation of the worm's 302 neurons would not have the information required to simulate the workings of the worm's 'brain': the organism's functional capabilities are far greater than is implied by its neural structure.

Multiply that complexity up a quarter of a billion times and you start to see some of the problems that uploaders face – though, in reality, things are more complex still, as the essential aspect of the brain's functioning comes not just from the individual neurons but from the network of connections between them – and each neuron can be connected to hundreds of others. It's also the case that the part of the brain primarily responsible for our conscious thought, the cerebral cortex, is extremely complex – unusually so – in a human. Although elephants, for example, have bigger brains containing more neurons than ours do, their cerebral cortex is significantly simpler. Overall, the complexity of the brain is unlikely to be accurately reproduced in a computer in decades and probably even centuries.

> **This means that any suggestion that you can come back to life is simply snake oil.**
>
> MICHAEL HENDRICKS
> neuroscientist (2015)

Over and above the massive connectivity of the human brain – which it seems hard to imagine we will ever be able to map in its entirety – the problem with an upload is that it does not involve a transfer of consciousness. Were it possible to upload a person's brain structure to a computer, that would presumably have to be done while the person was still alive. This makes it clear that what we have is not the uploading of the person, but rather a copying of their mind. Even if that copy were able to be conscious in the same sense as a brain (and we don't understand what consciousness is, so we certainly can't say that this will be the case), and if it were indistinguishable to others from the original you, your brain and its consciousness would still remain to live and die with your body. It's no more time travelling than sending a video of yourself to a different country is travelling through space.

Not everyone shares all of Hendricks' doubts. An Oxford University study suggested that the storage capacity for uploading may well be practicable at some point. Computer storage, once seen as a limiting resource, has expanded even faster than the growth of processing capability. In fact, it is the processing capability that the Oxford researchers were more doubtful about, because real-time emulation of a whole human brain is currently an unapproachable task, though they felt that it may be feasible in less than two centuries.

RELATIVE COMPLEXITY OF ANIMAL BRAINS

Species	Common name	Neurons	Cerebral cortex neurons
Caenorhabditis elegans	Roundworm	302	0
Felis catus	Cat	760 million	250 million
Canis familiaris	Dog	2 billion	600 million
Loxodonta africana	African elephant	257 billion	5.6 billion
Homo sapiens	Human	86 billion	16.8 billion

Most supporters of cryonics or uploading accept that their approach represents a long shot, but they use an equivalent of Pascal's wager to justify their support for these ideas. In the seventeenth century, French mathematician Blaise Pascal suggested that it was rational to believe in God. This was because the benefits if God did exist, in terms of eternal life, far outweighed the disbenefits of losing some time and wealth if he acted

> It appears feasible within the foreseeable future to store the full connectivity or even multistate compartment models of all neurons in the brain within the working memory of a large computing system.
>
> ANDERS SANDBERG AND NICK BOSTROM, Future of Humanity Institute (2008)

as if God existed, but it were not true. Similarly, the suggestion is that preservation into the future this way may be extremely unlikely to succeed – but it is being put up against a zero per cent chance of surviving if it is not tried.

Such approaches are far from our conception of time travel using a time machine – yet Einstein established the basic science to show that true time travel is entirely possible.

04 RELATIVITY OPENS UP TIME TRAVEL

'The whole development of the theory [of relativity] turns on the question of whether there are physically preferred states of motion in nature.'

ALBERT EINSTEIN (1923)

In the remarkable year of 1905, twenty-six-year-old Albert Einstein published four outstanding papers. At the time, Einstein had failed in his first attempts to get an academic position. Instead, he was working as a patent officer (third class) at the Swiss Patent Office in Bern, a job he managed to obtain through a friend of a friend's father. It was a role that did not challenge him (though he seems to have found it surprisingly interesting), which left him plenty of time to ponder his scientific interests and to produce what can only really be called amateur scientific papers. Yet what papers they were.

ALBERT EINSTEIN (1879–1955)

Einstein was born in a block of flats in the German city of Ulm. His early life was liable to upheaval as his father's attempts at business were rarely successful for long. Happy at home, the young Einstein enjoyed learning, but found the rigid education system frustrating. Left in Munich at the age of fifteen when the family business moved to Italy, Einstein rebelled, renouncing German citizenship and emigrating to Switzerland. On his second attempt, he got into the elite Zurich Polytechnic, but he was a lazy student and failed to get an academic post on graduating. While working at the Swiss Patent Office, Einstein published his first major papers in 1905, but it was not until 1909 that he got his first full academic position. In 1915, now based in Berlin, he published his masterpiece on the general theory of relativity, supplanting Newton's earlier gravitational work. Media coverage, cemented by his winning the Nobel Prize in Physics in 1921, gave Einstein international renown. As his fame spread, he made a number of visits to the UK and US. With increasing concerns for his safety under the Nazi regime, Einstein emigrated to America in 1933, taking up a position at the newly formed Institute for Advanced Study in Princeton, where he continued to work until his death aged seventy-six.

In one of these documents, Einstein used a calculation to establish the reality of molecules: perhaps surprisingly, at the time the existence of atoms and molecules as distinct physical entities was still in doubt. In a second paper, he explained the photoelectric effect where light falling on some materials generates an electrical current. In the process, he made it necessary that photons – particles of light – existed, fixing the foundations for quantum physics, an achievement that won him the Nobel Prize. And he showed that $E=mc^2$, which appeared (in a slightly different form) in a short extension to the paper that made time travel not just possible, but inevitable: his special theory of relativity.

Light and motion

Relativity has the reputation of being mathematically challenging – and that is certainly true of the special theory of relativity's big brother, the general theory (more on this in the next chapter). But it takes no more than high-school maths to demonstrate that special relativity makes inevitable the reality of time travel. And, perhaps surprisingly, the outcome is all down to the interplay of Newton's laws of motion and the unique behaviour of light.

One of Einstein's heroes was the Victorian Scottish physicist, James Clerk Maxwell. As we have seen, it was Maxwell who had identified what light was – an

> ❝ **Time has multiple, sort of parallel rates at which it flows, depending on the state of who's making the measurement and the state of who's in motion, and what conditions they are in.** ❞
>
> NEIL DEGRASSE TYSON (2017)

interplay between electricity and magnetism, which could only occur at one particular speed in any medium. (Light is fastest in a vacuum and slower in a substance, for example air, water or glass.) This was how Maxwell was able to identify light's nature, because he had calculated the speed that an electromagnetic wave would travel, and it turned out to be the same as the speed of light.

As a result, Einstein picked up on the strange reality that light, uniquely among moving natural phenomena, does not have a relative velocity. It's not particularly obvious that this will make time travel possible. But Einstein discovered that by plugging this absolute nature of light's motion into Newton's laws, three things happened. If an object is moving relative to wherever the observer is located, that observer will see time on the moving object passing at a slower rate, the object contracts in its direction of movement, and the object's mass increases. These outcomes are all fascinating, but from our viewpoint, it is the slowing of time, known as time dilation, that is the crucial element.

The time dilation effect

Somewhat counterintuitively, time dilation makes it possible to travel into the future by slowing time down. If we think of a spaceship going on a round-trip away from and back to Earth, and time dilation tells us that as a result of its movement, time on the spaceship slows down, then less time will pass by for the passengers on the spaceship than will be experienced by the people they left behind on the Earth. So, when the spaceship returns, the travellers will find that they have moved into the Earth's future.

You can envisage why time dilation happens as a result of light's constant speed by imagining a device called a light clock. This is a clock in which the 'tick' rate is controlled by the passage of a beam of light up and down between two parallel mirrors, one in the ceiling, one in the floor. Before we head into space, let's imagine we set up a light clock on Galileo's moving ship travelling smoothly along a straight stretch of water. As we saw in Chapter 1, Galileo realized that inside such a steadily moving ship, with no way of contacting the outside, it was impossible to detect that you are moving. So inside the ship, the light clock's beam will continue to bounce up and down from floor to ceiling and back exactly as it would when the ship wasn't moving.

> **Distort time and you open the barriers that prevent us from travelling to the future or the past.**
>
> JENNY RANDLES (2005)

However, let's imagine we can look into the ship from the shore. Let's say that from our viewpoint, the light starts off as it leaves the top mirror. In the time the light takes to reach the bottom mirror, the ship will have moved forwards. So instead of the beam of light heading straight downwards, it will travel at angle. Because the speed of light is very high, we don't usually notice this happening. But it will occur. And if the ship were moving fast enough – say, a tenth of the speed of light (it would have to be a very fast ship, which is the point where we need to move our experiment off the water and into space), the effect would be quite large.

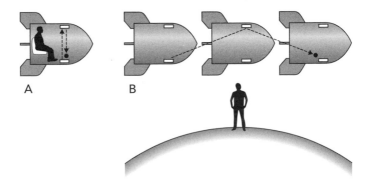

Now here's the thing. When the light travels at an angle, it has to travel further than taking the straight-line route up and down. If this happened with a normal object – a dropping ball, for example – this would not be a problem.

It's what we would expect because the speeds of the boat and the ball add together, producing just the right speed for the ball to make it down the longer distance to the bottom mirror at the correct time. But light's speed is *not* relative. It goes at the same rate despite the motion of the boat. So the only way that the light can arrive at the correct time from the viewpoint of us outside observers, after travelling further, is if time on the moving ship ticks by more slowly. The sheer act of motion has an impact on the passage of time.

Linking space and time

The impact of Einstein's special theory was to make space and time into an inseparable union. It's impossible to measure changes in one without considering changes in the other. (One of Einstein's former teachers, Hermann Minkowski, was the first to underline the move from having two separate concepts to considering a combined entity he called 'spacetime'.) Thanks to special relativity, movement through space destroys the concept of simultaneity – it is no longer possible to identify whether two events at separate locations are simultaneous. Depending on the motion of the observer, either event could happen before the other. Interestingly, Einstein illustrated this 'relativity of simultaneity' using an example of a pair of lightning flashes, observed from the ground and from a moving train. It seems that his work at the patent office, where he was involved in assessing a number of patents for electrical synchronization of railway clocks, had influenced his thinking.

Of particular importance to us, then, is that thanks to this linkage of space and time, special relativity shows that whenever we move, time slows down compared to the passage of time on a body with respect to which we are moving. The faster we go, the stronger the effect. As a result, by flying away from the Earth in a fast spaceship, because our time will have ticked by more slowly than that of a clock on the Earth, we return to find ourselves in Earth's future. This is not just an apparent difference – it is real and has been tested many times.

A wide range of experiments has demonstrated time travel into the future happening on a small scale. In the earliest, an atomic clock was booked onto a series of commercial plane flights to take it around the world (the scientists couldn't afford to charter a plane). When the clock returned home it showed a time a fraction of a second behind another atomic clock that did not move: the flying clock had moved into its future. In other experiments, measurements have been made of the lifetimes of short-lived

> ❝ Henceforth, space for itself and time for itself shall completely reduce to a mere shadow, and only some sort of union of the two shall preserve independence. ❞
>
> HERMANN MINKOWSKI
> (1908)

particles called muons, which are generated when high-speed incoming material from space called cosmic rays blasts into the atmosphere. The unstable muon particles live far longer than they otherwise would because their movement slows their flow of time as seen from Earth.

> " I can travel to the forward of your time. But I haven't yet figured out how to travel to the forward of mine. "
>
> JANNA LEVIN (2020)

As yet, the time machines we have constructed are very limited. Our best (if unintentional) time traveller to date is the probe Voyager 1, which was launched by NASA in the 1970s to observe the outer planets and is now headed out into the depths of space, still communicating with Earth. Voyager 1 has travelled about 1.1 seconds into the future in the course of its long-distance journey. To achieve significantly more we need to go a whole lot faster than we have to date, something that we will come back to in Chapter 8. The fastest a human being has ever travelled so far was on Apollo 10 way back in May 1969. The astronauts on board travelled at 39,896 kilometres (around 25,000 miles) per hour with respect to the Earth – which is only 0.000037 times the speed of light. If we want the kind of time travel that most of us envisage – moving months, years or even centuries into the future, then we need to get up to a far higher fraction of the speed of light.

Paradoxical twins

Assuming that such high speeds become possible, we could see the emergence of the easiest of the paradoxes of time to make real, known as the twins paradox. Imagine that there were two identical twins involved in a time-travel experiment – let's call them Lucy and Zoe. As is often the case with identical twins, their personalities are surprisingly different. Zoe wants to take on the universe; Lucy would rather stay at home. Zoe becomes a crew member on the first space flight to reach a sizeable percentage of the speed of light, travelling for five years before she returns to Earth.

When she left Earth, Zoe was thirty, so when she gets back home she is thirty-five. But she arrives back just in time to attend Lucy's fiftieth birthday party. This pair of identical twins, born the same day, are now years apart in age. This was possible because time ticked by much slower for the high-speed Zoe than it did for Lucy left behind here on Earth. While Zoe experienced five years passing by, Lucy waited for her sister to return for twenty years.

Zoe's experience emphasizes something crucial about real time travel into the future. First, it's not instantaneous as is usually portrayed in fiction. We can't just dematerialize in 2025 with a few special effects and reappear in 2045. A time traveller has to go on a spatial journey to undergo time travel, and that journey will take real elapsed time

that the traveller has to live through. Realistically, unless speeds that are very close to the speed of light can be achieved, any meaningful journey through time will take several years from the viewpoint of the time traveller. And during that journey, with a good enough telescope, the timeship would always be visible from Earth. A real time machine doesn't jump forwards through time: instead it produces a slowing of time within the ship.

If you have a logical mind, you may have spotted an apparent flaw in the whole special relativity time-travel trick. (For some reason, many people are obsessed with finding problems with relativity, though none of them has ever held up.) It is true that light's speed isn't relative – but the motion of the spaceship is. From Lucy's viewpoint on Earth, the spaceship is certainly moving. But from Zoe's viewpoint, the spaceship is stationary. (That's why Galilean relativity works.) For Zoe, it is the Earth that is moving away from the ship at high speed. And while the time on the ship slows as far as Lucy is concerned, from Zoe's viewpoint, it is time on Earth that slows down. For Zoe, time on the ship goes on perfectly normally. Remember, on the ship a light clock would continue to tick straight up and down. Zoe doesn't experience any slowing down of her time.

TIME DILATION BY NUMBERS

c is the speed of light in a vacuum: 299,792 kilometres per second.

Speed	Duration for traveller	Duration on Earth	Time travelled into future
0.1c	10 years	10.05 years	18.25 days
0.5c	10 years	11.55 years	565.75 days
0.9c	10 years	22.94 years	12.94 years
0.99c	10 years	70.88 years	60.88 years

This is all perfectly correct – and it might seem that the symmetry between the two sisters means that one can't age slower than the other, so when they are eventually reunited, they will both end up the same age. But that symmetry is an illusion. In reality, something happens to Zoe that doesn't happen to Lucy. The spaceship has a force applied to it to accelerate it away from the Earth up to near the speed of light. At the extreme of its journey, the ship's motors are activated again and the spaceship slows to a stop, turns around, then accelerates back to the Earth before finally slowing to a stop to be able to land. The force that causes the various accelerations is applied to the spaceship (and Zoe), but is not applied to the Earth (and Lucy).

Special relativity only applies to situations where there is no acceleration, which physicists call inertial frames. In such situations, things carry on moving as they always have. This is the case when the ship is moving away from Earth at constant speed. But when the ship's engines are engaged, it effectively resets the clocks, meaning that the ship does indeed arrive back to find more time has elapsed on Earth than it has onboard. The ship, and Zoe with it, has genuinely travelled into the future.

This is remarkable stuff. We now have the first essential for time travel: moving into the future faster than simply waiting for the time to pass. But, of course, moving forwards in time isn't enough for what is generally thought of as the complete picture of time travel: we want to journey into the past as well.

05 HOW TO VISIT THE PAST

'The past is a foreign country: they do things differently there.'

L. P. HARTLEY (1953)

One thing the designer of every science fiction time machine gets wrong is that forwards and backwards time travel are totally different. Moving into the future is easy – as we have seen, like Doc Brown's DeLorean, we just have to move at speed to achieve it. Admittedly, the 88 miles per hour speed required by the fictional time machine would only produce a tiny shift, but even a slow movement has some time-travel effect. However, getting into the past is much harder to achieve, requiring a totally different technology. Even so, physics tells us that it is, in principle, possible thanks to Einstein's masterpiece, the general theory of relativity.

Although it sounds as if it is a more general version of relativity theory than the special theory (which it is), crucially the general theory also explains the workings of gravity. The theory shows how matter warps space and time around it. We often concentrate on the space part – it's the warping of space by the sun's mass that means the Earth's straight-line motion through space takes it in an orbit around the sun, for example. But a gravitational field also twists time. With a powerful enough field, in special circumstances, this is sufficient to produce a kind of loop in time, enabling us to take a step into the past.

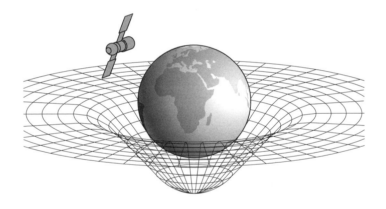

Einstein's happiest thought

Einstein claimed to have had his 'happiest thought', which started him working on the general theory of relativity, while sitting in a chair in the Swiss Patent Office. All of a sudden, he noted in a lecture he gave in 1922, 'a thought occurred to me. "If a person falls freely he will not feel his own weight." I was startled. This simple thought made a deep impression on me. It impelled me toward a theory of gravitation.'

This thought was the beginning of what became known as the equivalence principle. At the time, Einstein's idea was quite hard to envisage. Perhaps the best example available was the way that someone in a falling lift would not be pulled down against the floor of the lift by gravity as both the person and the lift would be accelerating at the same rate. Now, we have a more dramatic example: the International Space Station (ISS).

> **If the river of time can be bent into a pretzel, create whirlpools and fork into two rivers, then time travel cannot be ruled out.**
>
> MICHIO KAKU (2004)

We have all seen astronauts floating around onboard the ISS. A natural assumption might be that because the astronauts are up in space, they are not feeling the effects of the Earth's gravity. However, the orbit of the ISS is relatively low – just 350 kilometres (217 miles) above the Earth's surface. At that distance, the force of gravity from the Earth is nearly 90 per cent of that felt on the surface. What gives the astronauts the opportunity to float around is that they are in free fall, plummeting towards the Earth under its gravitational attraction – and so, as Einstein pointed out, they do not feel their own weight.

THE INTERNATIONAL SPACE STATION

There have been a number of space stations in the history of space flight, though none to rival the space stations of science fiction, such as the giant hotel-like rotating wheel featured in *2001: A Space Odyssey*. The earliest real stations were effectively large capsules that stayed in orbit, from the first, the USSR's Salyut 1, which spent 175 days in space in 1971, through to the USA's Skylab, in orbit between 1973 and 1979, though it was only occupied for a similar period. A true space station needs to be assembled from a number of modules to make it a long-term habitat. The first such station was the USSR's Mir, launched in 1986 and operational to 2001, but the most effective so far has been the Russian–American International Space Station, the first module of which went into space in 1998. Since 2000 it has continuously been occupied and enabled a range of experiments to be undertaken, though arguably its prime benefit has been to keep alive the public's interest in space. At the time of writing, 240 astronauts have visited this low-Earth-orbit facility.

When one body orbits another – whether it's the ISS orbiting the Earth or the Earth going around the sun – the orbiting body is falling towards the other one under the force of gravity. The only reason things don't end disastrously is that the satellite is also moving at 90 degrees to the direction of fall, just fast enough to stay at the same height. This is why for any particular distance there is only one speed the body can be travelling to stay in orbit. In the case of the ISS the speed is around 27,600 kilometres (17,150 miles) per hour. The astronauts are in free fall, but along with the space station itself they travel sideways at the right speed to keep missing.

What is happening here is that accelerating under the force of gravity cancels out weight. Einstein made the leap to suggest that acceleration and gravity are to all intents and purposes the same thing – indistinguishable in their effect. The acceleration part is why, despite being about gravity, we are dealing with the general theory of *relativity*. The familiar Galilean relativity tells us that in a closed vessel you can't tell the difference between the vessel moving steadily or being stopped. The special theory corrected this for its failure to deal with the tie-up of space and time. And Einstein's general theory brings in relative acceleration, telling us that in a closed vessel you can't tell the difference between the vessel accelerating or being exposed to gravity.

The equivalence principle

Think of what it feels like when a plane accelerates down the runway. The experience is one of being pushed back into your seat. In fact, what's happening is that the plane is accelerating forwards. As a result, the seat back pushes into you and, thanks to Newton's third law of motion, you feel an equal and opposite reaction of pushing into the seat. Now let's imagine being in a spaceship with some very precise measuring equipment onboard. The ship is accelerating. So just like on the plane, you will feel a force pushing you towards the back of the ship. If the ship is accelerating at the same rate as you feel from gravity on the surface of the Earth – around 9.81 metres (32 feet) per second every second – you will feel as if you have the same weight as you do on Earth.

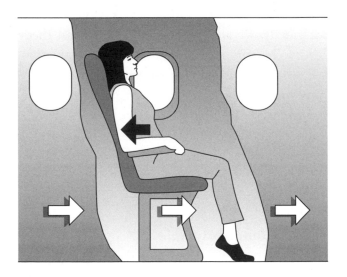

Now, let's send a beam of light across the inside of the ship. If the ship is accelerating, then during the time that it takes the light to cross its interior the ship will have moved slightly. As

> **Space tells matter how to move. Matter tells space how to curve.**
>
> JOHN WHEELER (1973)

a result, the beam would appear to bend slightly, rather than crossing in a straight line. But if acceleration is truly equivalent to gravity, then exactly the same thing should happen if the ship were sitting still on the Earth's surface. Not only would you feel your weight, but also the light's path would be bent by gravity.

Although the mathematics of the general theory is painfully complex (Einstein had to get help with the maths), the implications that massive objects cause space and time to bend – to be warped – can be appreciated without diving into the field equations of relativity.

The way, for example, that gravity causes the ISS or Earth to orbit is a large-scale version of the light crossing the interior of the spaceship. If we think of the Earth travelling around the sun, in reality the Earth is travelling in a straight line through space, as basic Newtonian physics would suggest. However, the gravitational effect of the massive star at the centre of the solar system is to warp space so that the straight line becomes curved around it. The result is that the planet moves in a curved orbit.

THE FIELD EQUATIONS OF GENERAL RELATIVITY

The general theory of relativity combines a number of gravitational effects to describe how matter and spacetime interact. This interaction is described by Einstein's deceptively simple-looking equation:

$$G_{\mu\nu} + \Lambda g_{\mu\nu} = (8\pi G/c^4) \, T_{\mu\nu}$$

This is beautifully compact, but uses a special notation that allows single symbols to stand in for a number of different equations. Each of the parts of the equation with the subscript $\mu\nu$ is actually a tensor, a mechanism that here is being used to collapse ten complex differential equations into this elegant form. Broadly speaking, however, the left-hand side describes the curvature of spacetime and the right-hand side the way that mass and energy provides a warp. Solving these equations is only possible for special cases. What this means is that the equations can be simplified by limiting what they apply to – for example, the first solution was for a perfectly spherical, homogeneous single object that doesn't rotate. But it isn't possible to solve the equations perfectly for most real-world systems.

It's not so obvious why a stationary object suspended above the Earth – Newton's famous apple, for example – begins to fall. However, we have to remember that general relativity tells us that matter warps space and time. If we imagine that the apple is moving steadily through time, if we warp this motion into another dimension it will also be moving through space (there's only one time dimension). Einstein's relativity always forces us to remember that space and time are not separate entities but part of the whole that is spacetime.

A twist in time

Although it's not mathematically exact, that picture of a movement in time being warped into movement through both space and time also gives us a first hint of why the general theory has the potential to make travel into the past possible. Think of moving through just two physical dimensions. Imagine that an ant is walking along the horizontal axis on a piece of graph paper. If the ant starts to walk on a curved path that moves towards the vertical axis then it will no longer be travelling as quickly along the horizontal axis. Similarly, when movement through time warps into movement through space and time, the

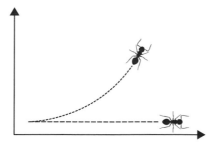

rate of travel along the time axis slows. Being exposed to gravity, like movement, slows time down.

It might seem at first glance that we've just found another way to travel into the future – but the situation is not the same. Just as Galilean relativity means we have to ask 'with respect to what?' when we talk about moving through space, we also need to ask 'with respect to what?' when thinking of travelling through time. Generally speaking, when we say we want to travel into the past, we mean the past with respect to the current time on Earth – so in itself, the fact that the Earth's gravitational field means that time runs a little slower on Earth than it otherwise would does not give us a route for travelling into the past. There is a way – but it's rather more complicated.

We'll come back to that in a moment, but it's worth emphasizing that just like the time-manipulating effects of special relativity, the influence of gravity on the passage of time has been widely tested, typically by using tall towers to cause an effective time difference because the gravitational field is slightly weaker at the top of the tower. In fact, the GPS satellites that support satellite navigation would not work if they weren't corrected for the influence of both special and general relativity.

Each GPS satellite is effectively a very accurate clock, constantly broadcasting a time signal. By measuring the difference between the times from a range of satellites, a receiver can work out where it is located. Because the satellites are moving, like the atomic clocks flown around the world, special relativity means that time ticks by a little slower on the satellites than on the Earth. But the satellites are also around 20,000 kilometres (12,500 miles) up – and so experience less warping of time by gravity than we do on the surface. This makes their clocks run a little faster. The general relativity result is the stronger of the two: if the satellites didn't correct for the combined effect, they would drift several kilometres away from being accurate in just a day, rendering the system useless. (Despite rumours that correction for this had to be available with an off-switch, because a US general didn't believe it could be true, the designers of the system always knew it would be necessary to correct for it.)

To travel into the past we need to discover or create a location where time is running slowly compared with our location – and to have a way to get into that location. This is why making a useful backward-travelling time machine is tricky. It doesn't break any laws of physics, but we need somehow to either jump from one place in space to

> **If you can bend space you can bend time also, and if you knew enough and could move faster than light you could travel backwards in time and exist in two places at once.**
>
> MARGARET ATWOOD (1988)

another or warp space sufficiently that we can effectively get into this slow-running location directly. We will explore more of what this means and how it can be achieved in Chapter 9, but the general theory of relativity has a range of interesting implications for space and time that might eventually make such an approach possible.

Harnessing the general theory

In total there seem to be at least three potential mechanisms to provide time travel into the past. Perhaps the simplest sounding is a Tipler cylinder, which involves creating an extremely massive cylinder (hypothetically most likely to be formed from a number of collapsed stars) that is rotated at high speed. This has the potential to be used as a time machine because of a well-tested aspect of the general theory of relativity known as frame dragging. What this means is that a rotating massive body pulls time and space around with it as it rotates.

A simple analogy is rotating a spoon in a jar of honey. As the spoon is twisted around, it pulls the adjacent honey into motion too. Because honey is viscous, the honey that is already moving will drag a little of the honey further

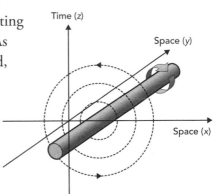

further from the spoon with it. The result is that, after a little while, there is a miniature honey whirlpool. Frame dragging means that a similar thing happens with spacetime around a massive rotating cylinder, providing the potential to link together two points in time.

A second possibility is to pick up on the observation that travelling faster than light would provide a mechanism to travel backwards in time. There is a considerable amount of theory devoted to the potential of creating a 'warp drive' for interstellar travel, like time travel, a staple of science fiction. We know that it is impossible to travel faster than light through space, but a warp drive enables a ship to move relatively slowly, but to have space itself warped around it so it gets from A to B faster than light. Although there are serious issues to be overcome, the basic design of such a drive, a so-called Alcubierre drive, has been published.

A final option involves the use of a wormhole, also known as an Einstein–Rosen bridge. This is a totally theoretical implication of the general theory of relativity. It suggests that it should be possible to link two remote points in spacetime by providing enough of a warp that one point is joined to another. It's another science fiction favourite to get around travelling interstellar distances in a manageable time, but it also provides a mechanism to – in principle – take a trip to the past.

One way to envisage this is that we set up a wormhole that links us to a distant spaceship. This spaceship has been travelling far and fast enough for time to have run slowly. So making the jump through the wormhole should effectively take us into the past. (We then have to get back, of course.) This might sound rather simpler than the kind of engineering required to manipulate stars. However, note that 'totally theoretical' in the previous paragraph. While frame dragging has been demonstrated, no one has ever seen a natural wormhole, nor has one ever been created. And even were it possible, there are serious concerns about their stability.

> **Wormholes, like any other possible form of travel faster than light, would allow one to travel into the past.**
>
> STEPHEN HAWKING (1988)

Creating a stable wormhole requires a conceptual capability known as negative energy. This exists for real in an observed physical phenomenon called the Casimir effect, but only on a tiny scale, making it difficult to see how it could be deployed. Still, there is a small hope here, one that was made use of to good effect in the science fiction movie *Looper*. How it may or may not work will be explored in Chapter 9.

One man particularly aware of the difficulties of backwards time travel is American physics professor Ronald Mallett, who has dedicated his life to researching time travel after his father died young. Although Mallett's inspiration was the thought of visiting his late father, he knows it would not be possible with his designs – he believes that he can construct a machine based on frame dragging that will enable particles to travel a tiny fraction of a second into the past, though using laser light rather than a Tipler cylinder to create the effect.

Why, though, don't we just wait for time travellers to come to us? After all, however far in the future people invent time machines, presumably, if it is possible to build them, at some point surely one will travel back in time to us and bring us the technology? So where are the tourists who travel through time?

06 WHY WE DON'T GET TIME TOURISTS

'One might hope therefore that as we advance in science and technology, we would eventually manage to build a time machine. But if so, why hasn't anyone come back from the future and told us how to do it?'

STEPHEN HAWKING (1988)

Professor Stephen Hawking cast doubt on the realities of backwards time travel in *A Brief History of Time*, asking why travellers from the future have not turned up. The neat thing about time travel is that it shouldn't matter when it's invented – so why aren't we swamped with time tourists?

This is something that science fiction has often pondered. Those who enjoy the idea of a dystopian future suggest that we will wipe ourselves out before we can develop a time machine. Those with more optimism believe that, rather like the 'Prime Directive' on *Star Trek*, which supposedly prevents Starfleet personnel from interfering with the development of external civilizations, future travellers have laws forbidding them to mess around with the past. Having said that, in *Star Trek* the directive seems often to be ignored, and it's hard to imagine that some individuals wouldn't still try to interfere with the past (hence the use of time travel in the movie *Looper* by criminal organizations). So perhaps the technology, or something fundamental in the mechanism of time travel, would prevent direct physical interaction.

Chronology protection and the time COP

It could be argued that we can't change the past simply because it is done and fixed. Rather like an old-fashioned photographic print, once it is taken the original reality can't be changed, even if the final print can be touched up. In such a universe, it wouldn't matter what you did: it would be impossible to change the past; yet by simply being there, a time traveller would change it. What's more, thanks to the butterfly effect, which says that in the chaotic systems of nature a small difference in starting conditions can result in huge changes in the future, even the smallest interference with the past could make a difference. However, such arguments tend to amount to little more than handwaving. We can say that we believe the past is fixed, but we can't prove that to be the case. And if the past were changed, it would simply be the past –

we wouldn't have a memory of an alternative past against which to compare it.

In 1992, Stephen Hawking published a paper in the reputable journal *Physical Review D* entitled 'Chronology Protection Conjecture'. Basing his argument on one of the more obscure mechanisms for backwards time travel, making use of a hypothetical, probably non-existent object called a cosmic string, Hawking appears to prove that the resultant 'back reaction' would prevent 'closed timelike curves' – the practical requirement for backwards time travel – from appearing, at least unless the cosmic strings used were infinite in length. Hawking concluded: 'These results strongly support the chronology protection conjecture: the laws of physics do not allow the appearance of closed timelike curves.'

However, this conclusion only applied to this very unlikely scenario.

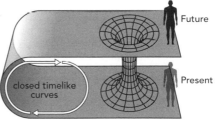

The same concept in a broader fashion is sometimes referred to as the Causal Ordering Postulate, primarily to get in a reference to the 'time COP'. All the problems and fascinating paradoxes of backwards time travel, which we will encounter in Chapter 10, derive from the disruption of causality and its implications.

Cause and effect

Causality is a crucial aspect of having a scientific understanding of the universe that is often misunderstood because of our built-in pattern recognition software. Human beings (and other animals) are very good at spotting patterns. It's a basic survival skill. If we had to learn every single experience anew, we wouldn't survive long. But with the broad pattern of what a predator lurking in the shadows looks like, for example, we can act quickly to enhance survival, even if we don't always get it right and sometimes run away from a shadow.

> **Were a time traveler from the future to access the internet of the past few years, they might have left once-prescient content that persists today.**
>
> ROBERT NEMIROFF AND TERESA WILSON, Michigan Tech University (2015)

Just how good we are at this pattern-spotting process can be deduced from work that has been done in the field of artificial intelligence (AI) on image recognition. Recognizing something in a picture is one of AI's greatest achievements. It does this better than almost anything else, and in a few special circumstances can beat humans. But here's the thing. To learn to recognize something, the software has to be trained with thousands of images. Humans can achieve similar levels of recognition after seeing a couple of examples.

The trouble is that we are so good at spotting patterns that we frequently do it when there is no true pattern to see.

In survival terms, it's better to err on the side of caution. But the result is that we often confuse correlation with causality. Correlation is where two or more things occur in proximity, whether spatially or in time, or both. Correlation certainly can occur when one event causes the other – but there is not necessarily a causal link.

Historically, the confusion of correlation with causality could result in someone being accused, say, of being a witch if there happened to be a cluster of bad events nearby. Unfortunately, random events do tend to come in clusters. (If you doubt this, think of emptying a box of ball-bearings onto the floor. It would look highly suspicious if they ended up evenly spread out. Instead you would expect to see some close together, others widely separated.) When such clusters of events happen near to a cause (a witch in the old days, a phone mast now, for example), there is a correlation, but there is not necessarily a causal link.

Where there is true causality – one event causing another – there is a cast-iron rule in everyday physics. The event that is the cause should come earlier in time than the event that is the effect. If effect comes before cause, then we've got our labels the wrong way round. Yet time travel makes it possible to disrupt that 'causal ordering' – hence the causal ordering postulate.

Even the easily demonstrated effects of relativity have the potential to disrupt event ordering, as Einstein discussed when looking at the impact of the special theory on the relativity of simultaneity. As we have seen, Einstein made use of the example of a railway train travelling along a track between two simultaneous events. These were a pair of lightning strikes that occurred at two widely separated locations on a long, straight piece of railway track. How could he check that the events were truly simultaneous? It's not possible to be in both locations at once.

Einstein suggested positioning an observer midway between the events (clearly this couldn't be done for real, as you would have to know where the lightning strikes would be, but it could be done with a pair of artificial flashes of light). If the two flashes arrive at the midpoint at the same time, we can say that these strikes were indeed simultaneous.

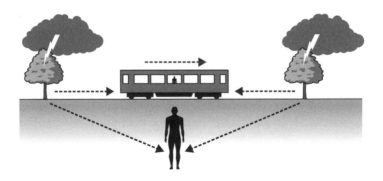

But now let's put our detector on a moving train that passes that midpoint at exactly the same time as the flashes reached it according to a fixed observer beside the track.

While the light beams are on their way, the train is in motion. As a result, the light will take longer to reach the detector from the flash behind the train than from the flash in front of the train. The events are no longer simultaneous. And which event comes first can simply be altered by having the train travel in the opposite direction.

> Time travel as a modern product, especially if allowed to the consumer, would probably soon come under some form of abuse.
>
> DAVID HATCHER CHILDRESS (1999)

Although this example demonstrates the feasibility of the changing ordering of events, it doesn't in practice change *causal* ordering, as one event is not able to cause the other, but the ability to travel backwards in time makes it entirely possible to change causality. Note that even if it is true, COP concept doesn't make time travel impossible, just the ability to change causal ordering. So, for example, it wouldn't make it impossible to travel into the future, or to go backwards in time if you could only see what was happening, but couldn't interfere in the past.

One suggestion for a mechanism for a time COP suggested by Hawking is dependent on an effect of quantum physics. One of the fundamental aspects of quantum theory is the uncertainty principle, which says that there are pairs of link properties where the more precisely one is known, the less precise we can be about the other. This is often used for the combination of position and momentum. The more precisely we know where a quantum particle is, the less certain we are about its momentum. It's a bit like taking a photograph of a moving car: very short exposure will capture the car in detail, but give no clue to its motion; a long exposure will leave the car a blur, but show how it's moving. But in some ways,

the more interesting possibility involves time and energy.

This pairing says that, even for apparently empty space, the more precisely we pin down a time interval, the less precise we can be about the amount of energy in that bit of space. Push the observation down to very short time intervals and the energy can fluctuate so widely that there is enough energy to generate matter. As $E=mc^2$ tells us, matter and energy are interchangeable, but it takes an awful lot of energy to make a small amount of matter (c^2 is the square of the speed of light – a very big number).

In such circumstances, pairs of particles – one matter, one antimatter (more on this in Chapter 8) – pop into existence, then recombine again, usually before they can ever be detected, though there are phenomena such as the Casimir effect (see Chapter 9) that demonstrate the existence of these 'virtual particles'. Hawking suggested that some such particles might be rendered real by the influence of a time machine, building up to the extent that their mass would warp spacetime sufficiently to undo the effect of the machine. This is very much an 'in principle' argument – there is no evidence that it actually would happen.

Where are they?

Even so, it still seems odd that we haven't been aware of time travellers if backwards time travel really is possible. (We wouldn't expect to see time travellers from the past, as this would require technology that we are not yet capable of producing to have already existed.) Back in 1950, the Nobel Prize-winning Italian physicist Enrico Fermi, while having lunch with co-workers at the Los Alamos National Laboratory in New Mexico, is said to have asked his colleagues, 'Where are they?' The topic of UFOs had recently been making news headlines in America: while not taking the news reports seriously, Fermi was asking a semi-serious question as to where the visitors from outer space were. It seems likely, surely, that we need to ask a similar question about visitors from a future time.

If we were going to see them anytime it should have been in 2005. In May of that year a convention for time travellers was held at the Massachusetts Institute of Technology. The idea was that people in the future would

> **We should breathe a sigh of relief. It means we were protected from the chaos that would result if someone came back and changed something.**
>
> DAVID BATCHELOR (2005)

know about the convention and would travel back in time to attend it. This was no small, backroom affair that would escape the notice of the press. Around 400 guests turned up at the elegant Morss Hall in Cambridge, Massachusetts – some hoping to meet themselves from the future and collectively hoping to ensure that the event would go down in the history books. There were musical performances and speeches. And at 10 p.m. on Saturday, 5 May, the hope was that the time travellers would arrive.

Sadly, no time travellers turned up. Did the organizers really expect a flood of exotic visitors from the future? Probably not – though they could always hope. Looking back on the event, it feels more like an entertaining university publicity stunt. The organizer, an MIT graduate student named Amal Dorai, claimed his inspiration was an internet comic strip called *Cat and Girl*. In the same vein, though less showy in its approach, earlier in that same year a plaque was erected in Perth, Australia, giving time travellers a place in space and time to rendezvous.

Like MIT's event, and an earlier, less polished time travel gathering organized by a group calling themselves

THE PERTH PLAQUE

In the event that the transportation of life from the future to the past is made possible, this site has been officially designated as a landmark for the return of inhabitants of the future to the present day.

Destination Day
12 Noon (UTC/GMT + 8 hours)
31st March 2005
Forrest Place, Perth 6000, Western Australia
Latitude – Longitude
31.9522 – 115.8591

We welcome and await you

the Krononauts, held in Baltimore, Maryland, on 9 March 1982, Perth seems to have had no visitors from the future.

Known for his sense of humour, Stephen Hawking followed in the footsteps of the 2005 events by throwing a time-travel party. Once again, no one turned up, but despite this, the day after the event he 'sent out invitations'. It shouldn't matter if it was the day after, as the invitations should still have been picked up in the future. Looking back at this now, it is relatively easy to understand why this kind of event did not have a chance of working. It's simple enough to find out that Hawking threw the party, but quite difficult to pin down exactly where and when.

Most reports put the date at 28 June 2009, but it has also been said that the party was in 2012. There are some articles identifying the location as Gonville and Caius College in Cambridge, UK, but the detailed invitation is hard to discover. According to another article, 'Hawking provided

> **Time travellers from the future should be pestering us with their cameras, asking us to pose for their picture albums.**
>
> MICHIO KAKU (2009)

precise GPS coordinates' – but given that it doesn't seem possible to locate these only a few years after the event, it's hard to believe that anyone will have them in, say, a hundred years' time. Similarly, the events and locations of 2005 are fading into the past now. It's entirely possible that no one in the future actually will have known about them (let's hope that this book will keep the knowledge alive!). But even so, it seems reasonable to ask in an echo of Fermi, where are they all?

Time-travel archaeology

One way to spot time travellers might be to look for traces of them online, which a pair of researchers at Michigan Technological University did in 2013. (Interestingly, Nemiroff and Wilson's paper gave the 2012 date for Hawking's party, adding to the confusion for poor travellers from the future.) The idea of the research was to spot anachronistic references to future knowledge or technology (much as there is occasional excitement in the tabloid press when someone apparently holding a modern mobile phone turns up in a period photograph).

The researchers made use of internet searches and social media posts, hunting for two specifics that would not be named until a well-known date: Comet ISON and Pope Francis. ISON did not gain the name until September 2012, so any references to it before that date could be evidence. Similarly, Jorge Mario Bergoglio, who become pope in 2013, chose a papal name that had not been used before, which made any earlier references to this of interest. Nothing turned up. The researchers also asked for time travellers to leave tweets on Twitter before the date of their request to do so, incorporating the hashtag #ICanChangeThePast2. (In case some time travellers believed, as some theorists do, that all attempts to change the past would fail, the option was also available to tweet #ICannotChangeThePast2. Again, nothing was received.)

It should be noted that even had something apparently prescient been discovered, this approach is not infallible. Coincidences happen. For example, in 1898, a novel named *The Wreck of the*

> **Even if it turns out that time travel is impossible, it is important that we understand why it is impossible.**
>
> STEPHEN HAWKING
> (2013)

Titan: Or, Futility was published. This described the sinking of a large British passenger liner called the *Titan*, considered unsinkable, which because of this did not have enough lifeboats for its passengers. In the book, the Titan sinks in the North Atlantic after hitting an iceberg. Clearly there are remarkable similarities with the sinking of the similar- sized *Titanic* in 1912. But there was no time-travel influence, and the reality is that there have been millions of novels published that didn't suggest any apparent time-travelling knowledge.

TOP FIVE TIME TRAVELLER DESTINATIONS

Fictional accounts of time travel often focus on a handful of events in history in which we might expect to detect a large influx of time travellers. Here's a top five:

Date	Event
Pre-65 million years before present	Seeing dinosaurs in the wild
Circa AD 30	The crucifixion of Jesus
22 November 1963	Assassination of John F. Kennedy
9 November 1989	Fall of the Berlin Wall
11 September 2001	Attack on the Twin Towers, New York

The idea of time-travelling tourists continues to provide entertainment, if not practical visitations. In February 2012, a blue plaque, similar to those put up by the UK's heritage industry, appeared on a building in Golden Square in London. It read 'Jacob von Hogflume, 1864–1909, Inventor of time travel, lived here in 2189'. Though the logic of the wording is a little misguided, it's a shame that the plaque, devised by Dave Askwith and Alex Normanton, didn't last long before being taken down.

In reality, there's a fundamental reason – one that Stephen Hawking should have realized in the first place (he later retracted his question). A time machine based on general relativity provides a gateway to a location where time slows down, taking us into the past – but such a gateway can never reach a point further in the past than when the time machine was first set up. There is no mechanism to travel earlier than when that link into the past was constructed. This means that if we want to echo science fiction and travel to see distant historical events or visit the dinosaurs, we have to hope some alien civilization started a time machine operating a long time ago – an unlikely prospect at best.

Such limitations apply to both data and people travelling through time, but there are extra ways information can move into the past that aren't available to human beings.

07 HOW TO SPEAK
TO THE PAST

**'Maybe to an American, Mozart's fortieth isn't
information.'**

GÜNTER NIMTZ (1995)

At a conference in Snowbird, Utah, in 1995, Austrian
engineer-turned-physicist Günter Nimtz set the cat
among the time-travelling pigeons. The topic of the
conference was superluminal transmission – the ability
to push light beyond the ultimate speed limit of around
300,000 kilometres (186,000 miles) per second. It
was thought this was only possible if no information
was transmitted. Nimtz produced his son's battered
Walkman and announced, 'Our colleagues assure us that
their experiments do not endanger causality. They say
that there is no possibility of sending a message faster
than light. But I would like you to listen to something,'
before playing a fuzzy version of Mozart's Symphony
No. 40. 'This Mozart', Nimtz announced, 'has travelled
at over four times the speed of light. I think that you

would accept that it forms a signal. A signal that moves backwards in time.'

The physics was unarguable. A signal sent faster than light can shift backwards in time and this recording did travel at more than four times light speed. (You can hear a copy of the recording at www.universeinsideyou.com/experiment7.html.) To this day, Nimtz's claims are subject to argument, but there are two other physical processes enabling instant communication through quantum entanglement and the generation of waves that travel backwards in time.

Frustrated total internal reflection

The mechanism behind Nimtz's high-speed music was a process known to Isaac Newton, though Newton was unable to explain it as there was no way to understand this effect, called frustrated total internal reflection, without quantum physics. Normal total internal reflection is the mechanism that allows, for example, a fibre-optic cable to send a beam of light bouncing along it without escaping. It occurs when a beam of light hits the boundary between two media at a shallow angle. If the light travels slower in the medium it's currently in than in the other and the angle is shallow, all of the light will bounce back.

A typical example is what happens in a glass prism. Usually, a beam of light passing through a prism continues out of the prism to the air. But light travels faster in air than glass – if it hits the boundary at a suitably shallow angle, it will bounce back into the prism. What Newton and others observed was that if a second prism is placed back to back with the first along the edge that was forming the boundary, instead of all the light being reflected some will pass on into the second prism. Somehow, adding a second prism that does not touch the first changes light's behaviour.

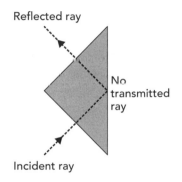

Reflected ray

No transmitted ray

Incident ray

Newton thought this was perhaps to do with parts of the prisms touching, but his being incorrect was not surprising as the actual cause is a phenomenon known as quantum tunnelling. Quantum particles, such as atoms and photons in light, do not have exact positions except at the moments they interact with other particles. All that

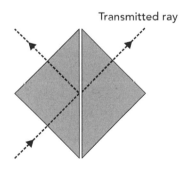

Transmitted ray

exists is a set of probabilities of being found in a range of locations. If there is a barrier that should prevent the progress of a particle, a quantum particle has a small probability of actually being on the other side of the

❛ **For the light which falls upon the farther surface of the first glass where the interval between the glasses is not above the ten hundred thousandth part of an inch, will go through that surface and through the air or *vacuum* between the glasses, and enter into the second glass.** ❜

ISAAC NEWTON (1704)

barrier. Though this is called quantum tunnelling, the name is misleading. The particle does not tunnel through the barrier. It appears on the other side without spending any time in the intervening space. (Recent research suggests tunnelling may involve a very small time interval, but not sufficient to make a difference to the outcome.)

Such tunnelling happens regularly. We use it in the flash memory that enables data to be kept without an electrical current in memory sticks and solid state drives – and it is essential for the operation of the sun. Our friendly neighbourhood star operates through nuclear fusion – turning smaller atoms into bigger ones, releasing energy in the process. To be precise, it operates on ions, atoms that are electrically charged due to losing electrons. Such ions repel each other strongly, so much so that the sun's gravity and pressure alone can't get them close enough to each other to undergo fusion. It's only possible because quantum tunnelling means that the ions can jump through the barrier of their own repulsion, getting close enough to fuse.

Although quantum tunnelling can overcome powerful barriers, it is only effective across short distances. Only when a pair of prisms are very close together will some of the photons in the beam of light jump across the gap and appear in the second prism. However, the interesting part is not so much the distance as the time it takes. As mentioned above, the particle spends no time in the barrier, simply appearing on the other side – and that's where the ability to beat the light speed limit comes in.

Imagine a photon crosses a certain distance in the first prism, then the same distance across the gap, then the same distance again in the second prism. Bearing in mind it takes no time to cross the gap, it will have crossed three units of distance in the time it should take to cover two – it will have travelled at 1.5 times its normal speed. In the case of Nimtz's experiment, photons travelled around four times the speed of light. This used microwaves and Perspex prisms rather than visible light and glass, while others used another tunnelling phenomenon in devices called undersized waveguides, but the principle is the same.

Will this ability to send a message backwards in time change the world? No. Not

> **Quantum tunnelling takes place all the time; in fact, it's the reason our sun shines.**
>
> JIM AL-KHALILI (2015)

all physicists accept that such superluminal experiments do involve genuine faster-than-light travel, as they suggest that it's more like a runner leaning forwards to hit the tape first in a race. The runner apparently travels faster, but this is due to a distortion of shape. Similarly, some argue that the wave-like nature of the photon is distorted, rather than it truly moving faster. But even those like Nimtz who are convinced that superluminal experiments involve information travelling faster than light accept that the time shift is so small that it would be impossible to use.

The abilities of superluminal experiments are the result of the strange behaviour of quantum particles – but these are not the only ways that quantum strangeness has the potential to play around with time. Another possibility emerges from one of the most fundamental quantum processes: entanglement.

> **No reasonable definition of reality could be expected to permit [the instant communication of quantum entanglement].**
>
> ALBERT EINSTEIN, BORIS PODOLSKY, NATHAN ROSEN (1935)

The Frankenstein effect

Quantum entanglement emerged from the last and greatest of Albert Einstein's attempts to undermine quantum physics. Einstein crops up a lot in the physics of time travel, but here he was attempting to do the opposite of what he achieved.

Initially, Einstein was a founder of quantum theory. The first to introduce the concept of quanta, Max Planck, only considered it a mathematical trick to make the numbers work and did not believe it implied anything about reality. Quantum theory depends on light, which up until then had been firmly accepted to be a wave, behaving as a stream of particles – photons. Einstein realized that if this were true, it would provide an explanation for a mystery of early twentieth-century physics: the photoelectric effect.

In this effect, the principle behind solar cells, incoming light knocks electrons out of a material, which start an electrical current flowing. If light were a wave, you would expect that by turning up the amplitude – making the light brighter – you would knock out more electrons. However, what actually happens is that some colours of light don't work, however bright they are. It's only high-energy light that triggers the photoelectric effect. So, for example, blue light might produce a current, but not red light, however bright it is.

Einstein showed that this would make sense if light were a stream of particles rather than a wave, and an individual particle of light was responsible for knocking out each electron. That way, the effect would be dependent on the energy of the light particle, not on how many of them there were. It was Einstein's paper explaining the photoelectric effect that won him the Nobel Prize. And with it he opened the Pandora's box of quantum theory.

EINSTEIN'S TOP QUANTUM PHYSICS QUOTES

Albert Einstein was the master of quotable remarks. Here are five of his top quotes arguing against quantum theory, mostly taken from letters to his friend, quantum theorist Max Born.

'I find the idea quite intolerable that an electron exposed to radiation should choose of its own free will, not only its moment to jump off but its direction. In that case, I would rather be a cobbler, or even an employee in a gaming house, than a physicist.'

'The theory says a lot, but does not really bring us any closer to the secret of the "old one". I, at any rate, am convinced that He is not playing at dice.'

'Quantum mechanics is certainly imposing. But an inner voice tells me that it is not yet the real thing.'

'The whole thing is rather sloppily thought out, and for this I must respectfully clip your ear.'

'This theory reminds me a little of the system of delusions of an exceedingly intelligent paranoiac, concocted of incoherent elements of thoughts.'

Unfortunately for Einstein, as the theory was developed, particularly by younger physicists such as Niels Bohr, Erwin Schrödinger, Werner Heisenberg and Max Born, it deviated from the pure, accurate reflection of the world that Einstein felt was essential for physics. Specifically, it showed that when a quantum particle was not interacting with another, it did not have a position, existing purely as probabilities of being in different locations until an interaction pinned it down. It was this probabilistic nature of quantum theory that Einstein hated and that resulted in his railing against the idea that 'God plays dice'. Somewhat like Frankenstein, attempting to destroy his own creation, Einstein would attempt to undermine quantum theory.

Spooky action at a distance

After a number of lesser attempts to find a flaw in quantum physics, usually presented as a challenge to Niels Bohr, which Bohr relatively easily defeated, in 1935 Einstein published a paper with the help of two younger physicists, Boris Podolsky and Nathan Rosen, that he felt demonstrated that quantum physics was fatally flawed.

The paper, 'Can Quantum-Mechanical Description of Physical Reality be Considered Complete?', is known as EPR after the initials of its authors. In it, the authors describe how a pair of particles produced together could be allowed to separate to a significant distance before one particle is observed. According to quantum theory, these particles would not have specific values for properties such as position or momentum or quantum spin until the property was measured for one of the particles. Yet the moment that happened, the laws of physics required the other particle to also have a fixed value for the property. Somehow, if quantum theory were correct, the information had to be transferred instantly from one particle to the other, however far apart they were. This, Einstein referred to as 'spooky action at a distance' (or *spükhafte Fernwirkungen* in German).

For Einstein, this proved that the quantum theory was flawed. As the paper triumphantly concludes, 'No reasonable definition of reality could be expected to permit this.' Either quantum theory was wrong or two things could communicate instantly at any distance. At the time this was not a real experiment. It would not be until the 1960s that experiments to test the outcome were devised. By the 1970s, the result was clear – and it remains the same to this day. Einstein was wrong. This instantaneous remote link between entangled quantum particles does happen.

COMMUNICATION TIMES AT LIGHT SPEED

Link	Distance	Time
New York to London	5,567 km	0.019 seconds
Earth to moon	384,400 km	1.28 seconds
Earth to Mars (closest)	54.6 million km	3 minutes, 2 seconds
Earth to Mars (most distant)	401 million km	22 minutes, 16 seconds
Earth to Jupiter (closest)	588 million km	32 minutes, 40 seconds
Earth to Jupiter (most distant)	968 million km	53 minutes, 46 seconds
Earth to Proxima Centauri (nearest star after the sun)	4.24 light years	4.24 years

Instant communication would be very useful. Communication delays can be irritating when using online meeting software, but also slow down the functioning of computers and make messages to other locations in the solar system difficult. For example, depending on the relative positions of the planets, a message can take a good twenty minutes to get from Earth to Mars. But from our viewpoint, instant communication is far more important, as it would enable a message to be sent backwards in time.

Chapter 4 showed that relativity makes forwards time travel possible. If a spaceship travels steadily away from Earth at high speed, time on the ship runs slow as seen from Earth. A message from Earth, sent instantaneously to the ship, would arrive before it was dispatched. But the situation is symmetrical. (Remember, it is only when the ship accelerates that the symmetry is broken.) So from the ship's viewpoint, time on Earth is running slow. This means that if the ship can relay that message instantly back to Earth it will arrive at Earth before it was first sent – it will travel backwards in time. Quantum entanglement can't help people travel back in time, but it looks like it could provide a mechanism to send messages into the past. Tomorrow's lottery results, anyone?

That's the dream. Now the reality. There's a problem. Despite decades of trying to come up with cunning work-arounds, no one has devised a mechanism to send a message using quantum entanglement. There are processes *involving* entanglement that send information – but they all have at least one stage using conventional light-speed communication. On its own, quantum entanglement does indeed communicate something instantly, but that 'something' cannot be controlled: it is entirely random.

To see why this is the case, think of the simplest entanglement, which is of a property of quantum particles called spin. (This being quantum physics, spin has nothing to do with rotating.) When the spin of a particle is measured, it can only have one of two values: up or down. If I keep one of an entangled pair and send the other to a distant location, when I check my particle and find its spin is in the up direction (described as 'spin up'), the other particle will instantly become spin down. It might seem that this spin value could be used to send a message in binary, which only requires two values, usually represented as zero or one. But I have no way of controlling the outcome. I can know in advance of time what the *probability* of getting up or down is – but I can't force a particular result.

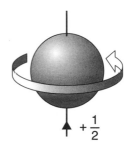

Local entangled particle

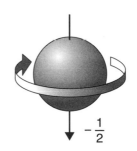

Remote entangled particle

This ability to transfer a random value is not useless. Such values are extremely useful for encryption of data, and entanglement is ideal to distribute an encryption key, as the values don't exist until a measurement is taken. Similarly, entanglement enables a process called quantum teleportation, which transfers the quantum properties of one particle to another – acting a bit like a miniature *Star Trek* transporter. This is very useful in quantum computers – but not for communication, as the process also requires a standard, light-speed message to be passed.

Waves that travel backwards in time

Entanglement, frustratingly, really is instant communication, but in a way that can't be used to send information. The other possibility relies on an oddity of electromagnetism that allows light to travel backwards in time. The origins of this oddity predate Einstein, going back to the work of one of his great heroes, Scottish physicist James Clerk Maxwell. As we saw in Chapter 1, it was Maxwell who worked out that light was an electromagnetic wave.

When Maxwell's equations are used to describe an electromagnetic wave, the equations have not one but two possible solutions. One of these was quickly swept under the carpet, because the 'advanced waves' it predicts travel backwards in time. According to the mathematics, when a light wave travels from A to B, an advanced wave travels from B to A, leaving B at the moment in time that the normal wave (known as the retarded wave) arrives and heading back through time to arrive at A at the moment the retarded wave departs.

Such strange waves had no place in Victorian physics – the advanced waves were simply ignored, even though there was no reason to do so other than the fact that they seemed bizarre. However, two leading lights of quantum theory, American physicists John Wheeler and Richard

Feynman, thought that there was a situation where advanced waves (or, at least, photons) were indirectly being detected.

Light is usually produced when an electron changes energy levels around an atom. An electron loses energy in the form of a photon of light. Photons have no mass, but they do have momentum, and when a photon is given off, the atom recoils, like a gun firing a bullet. Unfortunately, as part of the process, the atom's electromagnetic field acts on itself, causing a spiral of interaction that should produce infinite values.

Wheeler and Feynman suggested that rather than one, two photons were involved in the process. The ordinary photon left the atom, heading forwards in time, while the other started from wherever that first photon would eventually be absorbed, travelled backward in time and arrived at the atom to cause the recoil. This second, time-reversed photon was a quantum version of the advanced wave in action – and because this second photon caused the recoil, the model avoids the self-interaction that arose if the recoil came from the photon generated by the atom.

It was an elegant solution to a technical problem that was practically indistinguishable from

> **The law of interaction acts backward in time, as well as forward in time.**
>
> RICHARD FEYNMAN
> (1965)

what is observed. But it had one oddity. It only works if the photon produced by the atom is eventually absorbed. It doesn't matter how long it takes before that interaction, because the advanced wave photon travels backwards in time – but it has to occur eventually.

This opens a tiny window for an instant (and hence time-travelling) communication device. Let's imagine there was a direction in space where there was insufficient absorbing matter to stop all the photons sent that way. If Wheeler and Feynman's theory is true, a photon can only leave the original source if it will eventually be absorbed. So a tightly collimated beam of light – one that does not spread out widely as it travels through space – pointing in that direction will fail.

Now imagine taking a spacecraft off in that particular direction for some distance, and putting a huge absorbing blanket in place. Once the beam of light would have had time to reach the blanket, the beam will intensify, as all the photons in that direction are now being absorbed. If the distant spaceship now furls up and unfurls the blanket it will signal back to the source of the light beam. Each time the blanket is removed, the light beam will instantly reduce in energy – thanks to the advanced wave photons failing to travel backwards in time.

To turn this into a working information-based time machine requires a little more complexity – you would need to get an instantaneous message back in the other

direction, to a receiving station near Earth. And there is
the underlying assumption both that a theory that has
never been directly tested is correct, and that it is possible
to find directions in which a photon can travel for ever
without being absorbed by a distant piece of matter. But
it's still one way that it may be possible to get a message
into the past.

Whether or not this will ever happen, we know that
time travel into the future is achievable now. But what
would it take to make it useful?

08 WE NEED TO GO MUCH FASTER

'If everything seems under control, you're not going fast enough.'

APOCRYPHAL, ATTRIBUTED TO MARIO ANDRETTI

Although Voyager's 1.1-second journey into the future is impressive, it's not particularly useful. To travel years will require far more than Voyager's 61,000 kilometres (37,900 miles) per hour – and certainly more than the mere 39,896 kilometres (24,790 miles) per hour human speed record. To travel years into the future would require speeds more than ten thousand times greater.

Reaching such high speeds is not impossible – but requires a very different power source to today's chemical rockets. To understand the problems that rocket engineers face, we need to go back to basics. Almost all space propulsion depends ultimately on Newton's third law of motion, usually stated as 'every action has an equal and opposite reaction'. If you push something, it pushes back.

If you throw something away, it gives you as much oomph as you give it, in the opposite direction.

In search of reaction mass

Whether you're using a conventional chemical rocket or something more exotic, the chances are that your space propulsion will depend on pushing something out of the back of the spaceship – so-called 'reaction mass'.

This might seem obvious, but it was not clearly understood by many in the early days of rocketry. American rocket pioneer Robert Goddard was mocked by *The New York Times* when he proposed using rocketry in space. In a 1920 paper for the Smithsonian Institution, Goddard had proposed using a rocket to reach the moon. This attracted significant media attention, but *The New York Times* got it horribly wrong, commenting:

> That Professor Goddard, with his 'chair' in Clark College and the countenancing of the Smithsonian Institution, does not know the relation of action to reaction, and of the need to have something better than a vacuum against which to react – to say that would be absurd. Of course he only seems to lack the knowledge ladled out daily in high schools.

What the editorial writer assumed was that Newton's third law required something out there – air – to push against.

But in reality, the motor pushes the expelled fuel and the fuel pushes back on the motor, sending it (and the rocket) forwards. Such a motor needs plenty of fuel because it is the momentum of the exhaust from that fuel dispatched out the back that propels the ship forwards – and momentum is simply mass times velocity. You need to push out a significant mass to get up to a good speed. But until it is expelled, that mass is part of the spaceship – so a lot of energy is initially wasted on accelerating the yet-to-be-used fuel.

This is why the rockets we see carrying satellites and astronauts into space have multiple stages – it's pretty much the only way to carry enough fuel to get away from Earth, because of the need to get away from Earth's gravitation. However, even if a spaceship is refuelled when in space – pretty much an essential for a time-travel ship – it doesn't remove the problem of fuel. Or rather, the dual problem of fuel and reaction mass.

Fuel provides the energy to propel stuff out of the back of the rocket. Reaction mass is the stuff that gets shoved out of the back, providing the action and reaction predicted by Newton's third law. In a chemical rocket, fuel both generates energy and provides the reaction mass in the form of the gases produced when the fuel is burned. But with alternative forms of space motor, the two are totally separate.

Engage ion thrusters

Currently, the most common alternative to a chemical rocket is the ion thruster, which makes use of electromagnetic energy (this could be produced by anything from a battery to a nuclear source) to give the reaction mass its push. Here the reaction mass is made up of ions – charged particles of matter. Because the ions are electrically charged, they can be accelerated away from the spaceship using an electric field, generating thrust.

Such motors produce relatively low thrust, but can do so for a considerable time. As yet they have mostly been used for small navigational corrections, but with sufficient power and ionized reaction mass, they could gradually accelerate a ship over a lengthy period to a greater velocity than has yet been achieved. Ion drives like this can push out reaction mass at a higher velocity than a chemical rocket, meaning that it takes less mass to achieve the same effect.

Broadly, the problem of getting fast enough to make time dilation useful for time travel comes down to having sufficient energy. We know that energy is one of those things that the universe conserves. You may remember from school that the energy in a moving object is $\frac{1}{2}mv^2$,

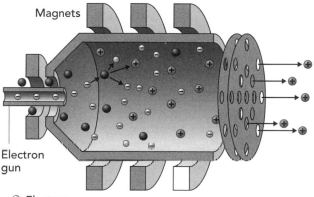

Magnets

Electron gun

⊖ Electron
● Neutral propellant atom
⊕ Positive ion

where m is the mass of the object and v its velocity. Let's do a bit of a back-of-an-envelope calculation on the energy needed to get a timeship up to a suitable speed. We're going to aim for $0.9c$ – 90 per cent of the speed of light – which would, for example, enable a time traveller experiencing an 8.7-year journey to return to find herself 20 years in the future – a time trip of 11.3 years. This calculation uses scientific notation where 10^n means 10 to the power of n (1 with n zeroes after it); so, for example, 1 million is 10^6.

The Dragon 2 capsule, used to take passengers to the International Space Station, has a mass of around 16,000 kg (35,000 lb) fully laden. $0.9c$ is 2.7×10^8 metres per second. Energy required ($\frac{1}{2}mv^2$) is $\frac{1}{2} \times 16,000 \times 2.7 \times 10^8 \times 2.7 \times 10^8$ joules = 5.83×10^{20} joules.

For comparison, total US electricity consumption is around 1.5×10^{16} joules a year. Our timeship would take the equivalent of 39,000 years of US electricity consumption.

In reality, the calculation above is generous because there's no such thing as a 100 per cent efficient motor. Not all the energy used up goes to propulsion – some of it will produce heat and vibration, for example. The engines on the Apollo missions' Saturn V rockets were between 6 and 12 per cent efficient – the rest was wasted. By contrast, ion thrusters can be as much as 80 per cent efficient, but still a little more will be needed than the simple calculation suggests for the timeship.

There's also a little problem that emerges from the scientific principle that is making time travel possible in the first place. The special theory of relativity does not only have an impact on the flow of time. It also influences the mass of an object. The faster something goes, the more mass it has. The Newtonian formulation for kinetic energy used above no longer applies at a sizeable proportion of the speed of light – it needs to be tweaked

> **It is important to realize that in physics today, we have no knowledge of what energy *is*.**
>
> RICHARD FEYNMAN (1961)

up. So our actual energy requirement will be more like
1.9×10^{21} joules.

Packing in the energy

At first sight, so much energy seems impossible to
generate – and it would be, using a conventional
rocket. The key factor here is energy density – how
much energy is packed into a fuel. The biggest chemical
rockets use liquid hydrogen. This packs in around 1.4
$\times 10^8$ joules for every kilogram of hydrogen. Only it's
not that easy. Some heat will be lost getting the fuel up
to temperature, but the bigger problem is that you can't
just burn hydrogen on its own. Combustion requires
oxygen. So you need to carry a greater mass of oxygen to
burn your hydrogen.

In the table below, the 'realistic mass' column includes
additional mass required for oxygen and lack of efficiency,
but note that this is only the fuel required to propel the
ship – not to move the fuel itself – nor does it include
extra energy to turn around and come back.

Kerosene – jet fuel – has an impressively high energy
density, significantly higher than that of the explosive
TNT (which is why the aircraft hitting the Twin Towers
in America in 2001 had such an impact). TNT is only
more dramatic because it burns faster. But clearly neither
kerosene nor liquid hydrogen are viable. The mass of
fuel required is vastly greater than the mass of the ship,

meaning that it would be impossible to get the entire mass up to speed.

COMPARING FIVE TIMESHIP FUELS				
Fuel	Energy density J/Kg	Simple mass for journey Kg	Realistic mass for journey Kg	Multiplier of mass of ship
Kerosene	4.3×10^7	4.4×10^{13}	3×10^{14}	19 billion
Liquid hydrogen	1.4×10^8	1.4×10^{13}	2.5×10^{14}	15 billion
Uranium	8.1×10^{13}	2.4×10^7	4.8×10^7	3000
Deuterium	5.8×10^{14}	3.3×10^6	6.6×10^6	400
Antimatter	1.8×10^{17}	1.1×10^3	4.4×10^3	0.36

A uranium reactor could be used to generate electricity for an ion drive, while deuterium – an isotope of hydrogen – could, in principle, pack in extra energy by using nuclear fusion, the process that powers the sun – though it is worth noting that despite fifty years of trying, we are yet to build a viable fusion power station.

Einstein's marvellous equation

In reality, the only realistic fuel is probably antimatter. Fans of *Star Trek* would be pleased, though there are no 'dilithium crystals' involved. Antimatter is a form of matter where the particles in the atoms have the opposite electrical charge to the usual one. Instead of electrons, antimatter has positively charged positrons. In the antimatter atomic nucleus, negative antiprotons replace positive protons. (To confuse matters, the neutron, which is electrically neutral, also has an antiparticle, which has opposite values of other properties.)

As readers of Dan Brown's *Angels and Demons* will know, when antimatter is brought into contact with ordinary matter, the mass of both particles is entirely converted into energy – a process known as annihilation – which can be used to power the ship. And because the equation that gives us the relationship of mass to energy is $E=mc^2$, a relatively small amount of antimatter produces a large amount of energy.

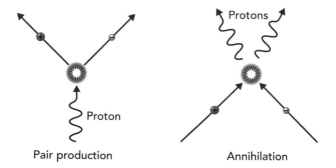

Pair production   Annihilation

Antimatter has been made, but here's where fact and fiction part company. The amount of antimatter produced is minute. Only a few millionths of a gram are made each year. The production techniques typically involve using high-power lasers to blast electrons into nuclei, or using particle accelerators to smash particles together. These collisions produce extremely high-energy photons, which can produce matter/antimatter pairs of particles.

> **If you see an antimatter version of yourself running towards you, think twice before embracing.**
>
> J. RICHARD GOTT
> (2001)

The antimatter particles need to be quickly separated and stored, as the moment they come back into contact with matter they will annihilate it. While it is perfectly possible to produce anti-atoms, the majority of antimatter captured is in the form of charged particles, because they can be trapped using electrical and magnetic fields to keep them away from touching matter. This kind of storage would have to be maintained onboard the ship.

There is no doubt that production of antimatter could be stepped up – much of the current production is a by-product, rather than something that is actively pursued. However, it would require a huge effort to produce the kind of quantities required – and bearing in mind how much energy would be released if it was allowed to escape, the security and safety protocols surrounding it would make nuclear weapons look trivial.

Things get even more complicated when we start to dig into the details. Antimatter produces energy in the form of photons – these would need to be converted into something that could be used to propel the ship, perhaps generating electrical energy (which would then need reaction mass to be accelerated by an electrical field). Alternatively, the light energy could heat propellant, which blasts out of the rear in conventional fashion.

The other big issue is that all the calculations above are for a one-way trip. We've looked at the energy required to get up to the appropriate speed. But there's not a lot of point setting up a time differential between a spaceship and the Earth if it never comes back. To do that, the ship would need to slow down to a stop, turn around and accelerate back up to near-light speed, then slow down again at the Earth.

If this process is done in the most wasteful fashion, we're talking about needing four times as much energy. However, there is a possible work-around, drawing on lessons learned from electric vehicles. Hybrid and electric cars use the kinetic energy from braking to generate electricity, recharging the batteries. Similarly, it should be possible to store away some of the kinetic energy of the timeship as it slows, reusing it on the return

journey. This is anything but
trivial – if, for example,
we are thinking about
storing it as antimatter,
there would need
to be a whole new
means of producing
the substance, as the
current approaches are too slow and take
up too much room. But in principle it could be done.

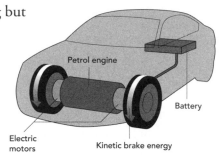

Getting away from carrying fuel

While we're considering mitigating factors, there are two
ways to reduce the mass of fuel to be carried. The first is
to make use of the fuel elsewhere and transmit the energy
to the ship. Transmission of energy was the dream of the
visionary (if somewhat unhinged) early twentieth-century
inventor, Nikola Tesla.

By far the biggest contribution Tesla made was in the
electrical engineering of alternating current, but he was
obsessed with (and lost a fortune as a result of his work
on) transmission of electrical power. It's simply impractical
over any range. However, we have a very good example of
transmission of electromagnetic energy in the sun.

Apart from a relatively small amount of energy from
the internal heat of the planet, all the energy that keeps
life going on the Earth comes from the sun in the form of

light – electromagnetic radiation. A spaceship can make use of light in two ways. A small amount of acceleration can be produced by light pressure – although light has no mass, it does have momentum and can give a ship a push. This would require huge sails. However, the main way light can be used is if those sails are also photoelectric cells, converting light energy into usable electricity.

A light-assisted ship would not be limited to the energy the sun could provide either. We have a means of creating a far more concentrated beam of light energy than sunlight in a laser. Lasers depend on an effect that was conceived by Albert Einstein – the stimulated emission of radiation. As we've seen, light is produced when an electron drops down an energy level in an atom. In a laser, photons first push up the energy of electrons, then the electrons produce extra photons as they are nudged into dropping down by further photons. The resultant beam is 'coherent' – this means that a property of the photons called phase is changing in step, making the light less likely to disperse.

By setting up huge lasers in space or on airless planets and moons, extra energy could be blasted to

> **Flying machines and ships propelled by electricity transmitted without wire will have ceased to be a wonder in ten years from now.**
>
> NIKOLA TESLA (1907)

> **The 100-gigawattt laser can be seen across the galaxy. It will be brighter than the sun.**
>
> PETER KLUPAR (2018)

accelerate the timeship, requiring no fuel on board. Although a relatively distant possibility, such technology is being considered seriously. In 2016, a $100 million technology grant programme was announced by a group called Breakthrough Initiatives to look into ways to accelerate tiny spacecraft to 0.2c using lasers and light sails. The aim is not time travel but getting these miniature probes to the nearby star Alpha Centauri in a journey taking perhaps twenty years. This Breakthrough Starshot programme could produce technology that would also be useful for time travel.

If a timeship is to rely on electrical energy produced from sunlight, though, it still needs ions to act as reaction mass – and this is where the second possibility to get around the need to carry masses of fuel comes in. Because space isn't empty.

There is free-floating gas and dust in space. With suitable scoops, charged to attract ions, it should be possible to collect some of the timeship's reaction mass as it goes, meaning that less needs to be carried. It's also possible with some kinds of propulsion system – fusion engines, for example – that fuel as well as reaction mass could be collected, as the most common material in space is hydrogen.

THE BUSSARD RAMJET

The best suggestion to date for a mechanism to collect both fuel and reaction mass from space, minimizing the need to carry extra mass, is the Bussard ramjet, dating back to 1960. Devised by physicist Robert Bussard, the idea is to use electromagnetism to attract charged hydrogen ions from space ahead of the ship. The Bussard mechanism would only kick in once the ship had undergone initial acceleration using conventional means. The motion of the ship through space would then be used to compress the incoming hydrogen to the extent that it would be relatively easy to start the nuclear fusion process, which would produce a constant flow of energy to power the ship, using its waste as the reaction mass. It's a neat idea, but in many regions of space there may not be sufficient hydrogen, and getting a fusion reaction going on Earth has proved difficult – it would be considerably harder in the restricted confinement of a spaceship.

Staying safe at speed

One final obstacle a timeship designer faces is navigating and staying safe at extreme speeds. Missing big objects like planets and moons isn't difficult – their positions are well enough known to make them easy to avoid. Even asteroids wouldn't be the problem that sci-fi movies like to portray. In a film, passing through an asteroid belt requires nerves of steel and instant reflexes as the spaceship dives around chunks of rock. In reality, our asteroid belt is far less crowded – asteroids are typically around 1 million kilometres (620,000 miles) apart: any particular route can be plotted safely.

The issue, rather, is the small stuff – exactly the same stuff that might provide free reaction mass. At these speeds, a particle of dust would blast through the hardest metal as if it weren't there. To make things worse, when matter collides with so much energy, it generates potentially deadly electromagnetic radiation. So our time travellers would need some kind of protective barrier. Ideally, this would be another science fiction favourite, a force field – but in the generic sense these do not exist. However, it is possible to use a strong electromagnetic field to repel particles – certainly something of this kind would be needed, along with conventional protective armour.

> **In fact, to get even close to an asteroid takes a great deal of effort.**
>
> BRIAN KOBERLEIN (2014)

The devil is in the detail with time travel based on the special theory of relativity. It certainly works, but making it practically useful is far from simple. At the time of writing there is renewed interest in manned missions to Mars. These feel achievable, but those who anticipate them any time soon overlook just how much more of a challenge it is than getting to the moon – taking perhaps six months, rather than the couple of days required for a lunar trip. This transforms the requirements both in terms of the supplies to keep the astronauts alive and the protection required against solar radiation. The requirements for a usable timeship are far greater than those of simply getting to Mars.

However, such difficulties are not insuperable in the long term. Humans have existed on Earth for about 200,000 years. We have been able to fly for around 120 of those years and have been able to reach another body in space for fifty years or so. Technology moves on and so will our capabilities. It would be significantly easier to send a small probe into the future for experimental purposes, though it would be of limited benefit other than testing the technology, because anyone can send an object into the future simply by hiding it away to be discovered later. Time capsules are, in effect, probes into the future. It is only when humans can make the journey that the effort becomes worthwhile.

The possibilities for what might await such time travellers are endless – though we have to bear in mind that, unlike their science-fiction equivalents, there would be no coming back from the future. At least, not until we can conquer the far greater challenges of interstellar engineering.

09 WE NEED A BIGGER TIME MACHINE

'We thus have experimental evidence both that spacetime can be warped ... and that it can be curved in the way necessary to allow time travel.'

STEPHEN HAWKING (1988)

We have seen that the difficulties of achieving backwards time travel are vastly more complex than moving forwards. However, even visiting the past may not be impossible for ever. Most ideas for doing this require technology that is thousands or even millions of years into the future – but Ronald Mallett, who we met briefly in Chapter 5, believes that it should be possible to demonstrate a reverse time-travel effect on a tiny scale in the laboratory now.

Time travel on the workbench
Inspired by a comic book version of H. G. Wells' *The Time Machine*, Mallett set out as a teenager to gain the

expertise necessary to build a time machine, in the hope of seeing his late father once more, who died when Mallett was ten.

In the 1970s, when Mallett was at the early stages of his career, openly admitting to an interest in time travel was career-limiting. Even Stephen Hawking was initially wary of bringing up the subject. Mallett worked first on lasers and then moved on to the general theory of relativity, which, as we have seen, is at the heart of moving backwards in time.

By the time he was qualified, Mallett was already aware that it would be impossible to build a time machine to go back to the 1950s, but this didn't dampen that initial drive and he continues to push for the development of an experimental time-travel device, though media exposure has meant that he now probably spends more time talking about his ideas than on practical work.

I tried to build the machine exactly as I saw it on the cover of *Classics Illustrated*, using television tubes, discarded pipes and other junk.

RONALD MALLETT (2009)

Mallett's key idea – still controversial – was that while it would not be practical to use the spacetime warping effects of massive bodies in the lab, it should be possible to achieve a much smaller similar effect using rings of laser light. He has published details of theoretical devices where laser beams are sent around tight loops forming a kind of time tunnel. The reason that in principle these could work is that light itself distorts spacetime.

RONALD MALLETT (1945–)

Mallett was born in Pennsylvania and brought up in New York. After a time in the Air Force, Mallett attended Penn State University where his PhD was on general relativity, specifically time reversal in a particular kind of hypothetical universe with curved space. He worked on lasers in industry before moving to the University of Connecticut as an assistant professor of physics, a post that was still rare for an African American scientist at the time. In 1998, he realized the significance of an aspect of general relativity stating that light can generate a gravitational field. With his expertise in lasers, he was able to conceive of rotating rings of light producing a frame-dragging effect. Although he continued to work on physics that supported his goal, he was not explicit about his interest in time travel until shortly before publishing his book, *The Time Traveller*, co-written with Bruce Henderson. Now an emeritus professor at Connecticut, Mallett continues to pursue his goal of time travel.

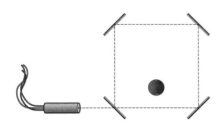

Although light does not have mass, it has energy, and that is enough to warp space and time. The effect is tiny. Even with Mallett's idea of a high tower of spiralling laser light, a stream of particles passing down the tower would only arrive at a time that was fractionally different to that expected. Designing such an experiment was fraught with difficulties. Light doesn't naturally travel in a spiral, but putting it through fibre optics would not produce the desired effect.

In the end, aided by experimental physicist Chandra Roychoudhuri, Mallett devised a setup where two thousand ring lasers, each reflected around the sides of a square, could produce a measurable effect on a particle passed through the tower as a result of frame dragging. Although this experiment was already devised in the first decade of the twenty-first century, at the time of writing it has not been attempted, due to lack of funding and Mallett's retirement.

Other physicists are sceptical, both about the ability to produce an experiment that duplicates the conditions in Mallett's theoretical setup and as to whether or not the frame-dragging effect would be strong enough to be detectable. Ideally, the answer would be an experiment, but whether or not this will be funded is unclear.

Factoring in the warp drive

Strangely, the most likely possibility for large-scale backwards time travel may come from another science fiction favourite, the warp drive. Usually, SF stories keep space journeys and time travel separate. However, time and space are not separable entities, but an intertwined whole in spacetime. As a ship approaches the speed of light, time slows to a standstill. Any kind of faster-than-light space travel is also inherently a time machine – the TARDIS remains one of the few fictional devices where this is explicit.

Variants of a warp drive have turned up in science fiction since the 1930s, and became widely familiar from the *Star Trek* TV show. The concept is rooted in general relativity. You can't travel faster than light, but there's nothing to stop space itself being warped at any speed – you can imagine a ship that is effectively stationary, but with the space in front of it being squashed up and the space behind being stretched – the result is that the ship gets to a new location without breaching the light-speed barrier.

> I can't believe that in all of human history, we'll never ever be able to go beyond the speed of light to reach where we want to go.
>
> WESLEY CLARK (2003)

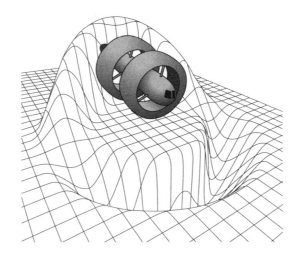

It's a neat idea, but one that was generally considered a work of fiction, at least until 1994. It was then that theoretical physicist Miguel Alcubierre dreamed up what was known for some time as the Alcubierre drive, though it tends now just to be called the warp drive. Let's be clear – Alcubierre didn't *make* a warp drive, but he took us from having no idea how to create such a drive, to knowing how it could be theoretically done, despite not being able to construct it yet.

Alcubierre was based in Cardiff, Wales, but the idea then moved to the natural home of space travel, NASA. It was from one of the organization's more extreme research outposts in 2012 that physicist Harold White would come up with a high-level design for a real warp drive. The original Alcubierre concept would have required a vast amount of energy to power it, but White's modification made it, at least in principle, possible to travel at warp speed using the energy from less than a tonne of antimatter.

Getting sufficient energy to power the ship was not the only issue, however. To function, the warp drive would also require a source of negative energy. As we saw in Chapter 5, although negative energy sounds implausible, there are real-life examples, notably the Casimir effect. This is observed when two flat metal plates are brought extremely close together without actually touching. The plates are attracted to each other, despite no magnetic involvement.

As we have seen, quantum theory shows that virtual particles very briefly pop into existence and disappear again due to fluctuations in the energy of empty space. In the Casimir effect, because the plates are very close together, there isn't room for many particles to pop into existence between them. More particles, however, briefly materialize outside the plates. Here, some will bump into the plates before disappearing again. The result is a pressure from the outside, producing a form of negative energy between the plates.

As yet there is no way to make sufficient negative energy to be useful, nor is there a way to deploy it – and it may never be possible – but at least the warp drive shows a small potential for a faster-than-light travel mechanism, and hence backwards time travel. NASA has made serious explorations of the possibilities of a warp drive – but this is not the only long-term possibility.

Pulling the strings

Should we ever reach the stars and have the capability to interact with large-scale galactic phenomena, two possibilities for backwards time travel emerge in the form of cosmic strings and Tipler cylinders. The cylinders, which we'll return to in a moment, have the benefit of being dependent on something we know to exist – neutron stars – whereas proponents of cosmic strings are building speculation on speculation as there is no evidence that cosmic strings are real.

Cosmic strings have no direct connection with the speculative 'theory of everything' known as string theory, although there has been an attempt to bring them into this model. Instead, cosmic strings are quantum structures that a theorist in the 1970s suggested could have formed in the early universe as the different forces of nature split off from each other. A cosmic string would be both incredibly thin – around a quadrillionth of a metre (25 trillionths of an inch) across – and vast in length. (Unless it were a loop, it would have to be infinitely long.) Cosmic strings would not be made of anything, but rather would be a distortion in the gravitational field acting as if it were incredibly dense – a metre of cosmic string would have a mass of a million trillion tonnes.

If cosmic strings existed (bear in mind they are mathematical concepts based on no evidence) then, in principle, if you could arrange for a pair of the strings to move away from each other at near to the speed of light and then managed to fly around them, the distortion the strings produced in spacetime should be sufficient to set up a 'closed timelike loop' – a path to an earlier time.

It might seem that such hypothetical concepts, driven by mathematics, are the playthings of mathematicians with no connection to the real world. However, in the last forty years, a large number of theoretical physicists have spent their time working on models that have a tenuous link to reality. Many, for

> **Anti-de Sitter space is a space with a negative cosmological constant. It is popular among string theorists because they know how to make calculations in this space. Trouble is, the cosmological constant in our universe is positive.**
>
> SABINE HOSSENFELDER (2019)

example, work on model universes where fundamental characteristics of the universe are different from the real one – because the mathematics works more effectively.

Cosmic strings are an example of this kind of highly speculative physics. The Tipler cylinder, by contrast, is more grounded in reality and provides a good example of a solution to backwards time travel that is limited by engineering capabilities rather than fundamental physics.

Cosmic engineering

As we have seen, backwards time travel is intimately associated with Einstein's general theory of relativity. The equations of general relativity are impossible to solve for a general case, but can be solved for a specific, relatively simple object. The first important solution in 1915 was for what would later become known as a black hole. But around twenty years later, the Dutch physicist Willem van Stockum came up with the solution for an infinitely long rotating cylinder of dust. If such a cylinder rotated quickly enough, the twist it applied to spacetime would be sufficient to be able to travel backwards in time by flying around it (as always, limited at the extreme to the point in time the cylinder started rotating).

Clearly, like the cosmic strings, this is nothing more than a toy concept, a thought experiment. But a variant dreamed up by American physicist Frank Tipler in the 1970s would present a mechanism that could, in principle, be made real.

Like Mallett's lasers and van Stockum's infinite cylinder of dust, a Tipler cylinder relies on frame dragging to pull space and time around with it. To make this mechanism effective, though, requires very concentrated mass: a material far more dense than anything we have ever directly experienced. The densest substance we encounter on Earth is the element osmium. One teaspoonful of osmium has a mass of over 100 grams (3.5 ounces). But to make Tipler's concept a reality would require a substance that was trillions of times more dense. This sounds just as fictional as van Stockum's dust cylinder, but we have good reason to expect that such a substance exists, in neutron stars.

Almost all of an atom is empty space. Electromagnetism and quantum physics make it impossible to do away with this. But one of the particles in the atomic nucleus, the neutron, has no electrical charge, meaning that a body of any size could be made up of neutrons crammed together. Assembling such an object by hand would be impossible – but nature has found a way. When a type of white dwarf star becomes unstable (typically by absorbing material from a companion star) it can undergo a massive explosion called a supernova. The outer layers of the star are blown off, compressing its core to such an extent that all that is left is a ball of neutrons. So dense is a neutron star that a teaspoonful would have a mass of around 100 million tonnes.

Neutron stars, therefore, form the ideal building blocks for a Tipler cylinder. However, we shouldn't underestimate the challenge of creating one. It would require around ten or more neutron stars to be pulled through space and assembled into a single body. Although neutron stars do mostly rotate quickly, the chances are they would be rotating in different directions, requiring significant manipulation. There would also be a problem that by bringing the neutron stars together, we would tip the whole over the limit where, unrestrained, they would collapse together to form a black hole.

We know where a good number of neutron stars are. The nearest so far are about 400 light years away. Without a warp drive this would mean it would take over 400

THE FIVE NEAREST KNOWN NEUTRON STARS		
Name	Location (direction)	Distance (light years, approx.)
RX J1856.5–3754	Corona Australis	400
PSR J0108–1431	Cetus	425
1RXS J141256.0+792204 (Calvera)	Ursa Major	625
PSR J2144-933	Grus	600
RX J0720.4-3125	Canis Major	1000

years to reach them. (And if we did have a warp drive, we wouldn't need a Tipler cylinder.) So the task of forming such a cylinder is not just difficult, it is ridiculously difficult. But the difference from the cosmic string concept is that while we certainly can't make a Tipler cylinder in the foreseeable future, there is nothing about it that is strictly impossible. This makes the Tipler cylinder rather better than the remaining significant contender: to make use of another science fiction favourite, the wormhole.

Alice through the wormhole

Like the rest of our backwards time-travel toolkit, wormholes are conceptual products of the general theory of relativity. As we have seen, wormholes are rips in the fabric of spacetime linking two points that otherwise would be much further apart. The 'wormhole' name reflects the way that a wormhole through, say, an apple provides a shortcut through space compared with having to travel around the outside of the apple, though invoking general relativity means they link locations not just in space, but in spacetime.

The alternative title of Einstein–Rosen bridge reflects the publication of joint work by Einstein and physicist Nathan Rosen in the 1930s and it forms an extension of the concept of the black hole. As we have seen, black holes were the result of the earliest special-case solution of Einstein's gravitational field equations and can be thought of as matter that has collapsed to such an extent that it effectively disappears to a point.

As you get closer and closer to a body the gravitational pull increases. Because there is no stuff to get in the way, you can get so close to a black hole that the gravitational field warps spacetime so much that even light can't escape. The distance at which this occurs, known as the Schwarzschild radius after the German physicist who first solved the equations, forms the black hole's 'event horizon'. There is no physical barrier at this point, though: something passing the event horizon wouldn't notice it.

Schwarzschild never imagined a black hole could form – there was no known way to compress matter sufficiently to achieve it – but as understanding grew of the structure of stars it was realized that it might be possible for a star with sufficient mass to collapse so dramatically as it neared the end of its life that nothing could resist the gravitational attraction of its component particles.

For decades, the reality of black holes was uncertain – they existed in theory but had not been detected – but now there are a good number of bodies, detected by their influence on matter and light around them, that are considered to be black holes. From the outside, a black hole is a perfect sphere. But the distortion of spacetime it produces is such that it can be considered internally as a kind of funnel that becomes narrower and narrower heading off to infinity. What Einstein and Rosen pondered was the possibility of having two such funnels intersecting.

If we imagine heading into one funnel, it seems reasonable that we could emerge from the other funnel at a different point in space and time. However, there are difficulties. It is perfectly possible to approach a black hole and come away from it again, provided you don't pass through the event horizon – but if you do pass that point there is no coming out, and the existence of a bridge into another black hole doesn't help.

It's not that passing through the event horizon is necessarily a subjective problem. Get too close to a black hole and a traveller will certainly have issues. The difference in gravitational pull between the traveller's closest point and their furthest would lead to them being stretched, a process graphically known as spaghettification. But for a large black hole, the horizon will be passed before this becomes an issue, and the traveller would never notice. To be able to exit from the second part of the wormhole, though, would require the traveller to be outside that black hole's event horizon.

Some have suggested that, rather than a pair of black holes, we need a black hole–white hole combination. A white hole has the reverse characteristics of a black hole – rather than keeping everything in, it spews everything out. However, the big problem here is that there is only evidence for one such entity: the universe at the point of the Big Bang. Which isn't particularly helpful.

So let's stick with the basic wormhole. Even if we can overcome the problems we've already met, there's another one. A wormhole should collapse if something attempts to pass through it. The collapse would occur so quickly that it would not be possible to get to the other end. However, there is a sort of solution. We need to hold the wormhole open – and that requires negative gravitational force. Something that would be generated by the kind of exotic negative energy likely to be required for a warp drive.

> **What one needs, in order to warp spacetime in a way that will allow travel into the past, is matter with negative energy density.**
>
> STEPHEN HAWKING (1988)

It is also, of course, not enough that the wormhole should exist and be traversable. The time at the far end needs to be behind the time at our end. The easiest way to make this happen would be if we could pick up the far end and swing it around very quickly, building up a special relativity time differential. Or you could put the far end near a neutron star, so the gravitational effect of slowing time would be brought to bear. But when you consider we don't know how to find or make a wormhole, performing this feat seems unlikely. Oh, and having got to the other end you need to get back. There's no point being in our past but fifty light years away (or whatever). And there's no point either just coming back through the wormhole, as that would involve travelling back to 'our' time. So you would need two wormholes with time differentials in opposite directions.

It's also worth saying that although wormholes are based on more solid theory than cosmic strings – we know that objects appearing to be black holes exist – we have no evidence that there are wormholes. They have never been detected naturally and we don't know how to make them. It is hopefully clear by now that a Tipler cylinder is less complex, while a warp drive holds out a bigger hope. Wormholes are fun, but unlikely to provide the solution.

Should backwards time travel ever be realized, though, even on the tiny scale of Mallett's hypothetical experiment, the paradoxes of time would begin to open up.

10 PARADOXICAL POSSIBILITIES ENSUE

'One should not think slightingly of the paradoxical, for the paradox is the source of a thinker's passion ... The supreme paradox of all thought is the attempt to discover something that thought cannot think.'

SØREN KIERKEGAARD (1844)

A number of paradoxes emerge if backwards time travel becomes real. There is often confusion about what is meant by a paradox – some use the term to mean a fallacy, an outcome that is dependent on a logical error, but it works better as a word for something that appears to be impossible, yet still happens. The paradoxes that the traveller who heads back in time uncovers result from the disruption of the simple flow of cause and effect. The outcome is fascinating. We know that physics appears to allow backwards time travel, yet were it to occur, it seems possible that we could get into a state where reality contradicts itself.

The useless machine radio

Even with a message heading a fraction of a second back through time, something we could envisage making possible with Ronald Mallett's design, we could ease gently into the nature of a time-travelling paradox. Imagine a radio transceiver that can be turned off using a radio signal – a perfectly possible concept. And let's imagine that the transceiver can itself generate the signal that turns itself off. Again perfectly feasible.

It then becomes one of those pointless but entertaining devices, typified by Marvin Minsky's 'useless machine'.

Marvin Minsky was an American computer scientist based at MIT, who specialized in artificial intelligence. He made what he called the ultimate machine, a box with a switch on it. When the switch was thrown, a mechanical pushing device emerged from the box and turned the switch off again. It was a device with the single purpose of turning itself off – and our radio transceiver is a non-mechanical version of the same concept. To keep it simple, we'll say that immediately after turning the device on, it is programmed to broadcast a signal that it can itself receive, which turns it off again.

Let's imagine we had a way to send that signal back one second in time (the displacement could, of course, be significantly less). We switch on the transceiver. It sends the signal, which travels back to one second before the signal was sent and turns the device off. Because the transceiver was switched off before the signal had been sent, the signal would not be sent back in time. But because the signal wasn't sent back in time, the transceiver was still on and the signal would be sent. But if that was the case, the transceiver would have been turned off before the signal was sent ... and so on.

> **Even a god cannot change the past.**
>
> AGATHON (born circa 445 BC)

This typifies the baffling, mind-tangling paradoxes of backwards time travel, though they can be much more dramatic.

What have you got against grandfathers?

Rather in the same way that the grand magic tricks of large-scale magicians who make cars or buildings disappear make use of the same techniques as the illusions of a table-top magician, only seeming more dramatic because of their impact, so what is probably the best-known of the time-travel paradoxes relies on exactly the same concept as the useless radio, but hits us between the eyes because of the scale of its implications.

Known as the grandfather paradox, it involves a traveller heading back in time and killing his or her own grandfather before the grandfather's children were conceived. However, if the protagonist's grandfather didn't exist, causing his or her mother or father not to exist, then our time-travelling murderer can't have been born either. Which means the murder was never committed. Which means the time traveller would be born after all ... and so on.

Although it is universally known as the grandfather paradox, I can't find any justification for it being a grandparent that is killed. The concept appears to date back in fiction at least to the 1920s, but why the archetype for this kind of time-travel paradox is usually a grandparent, rather than a parent, killed before the protagonist's conception, is not entirely clear. Perhaps it was just that a young version of a grandparent is more of a stranger than a parent, making it easier for our killer to carry out their unpleasant experiment.

By its own bootstraps

Even more baffling (and certainly less stressful) than the grandfather paradox are bootstrap paradoxes (a reference to 'pulling yourself up by your own bootstraps') where passing an object through time can make something appear from nowhere. What if, for example, we were able to use a time machine to send a book into the past before it was written? We send it back to the author who, being lazy like most authors, decides, instead of writing the book, to simply copy it out word for word and send the manuscript to the publisher. Who then wrote the book? Where did it come from?

If we think through the experiences of the two versions of the author – the 'present' and the 'past' author, we can see that there has to be some kind of disruption in the timeline. At the start of the experiment, the past author wrote the book and the present author will remember doing so. The present author then sends the book back to before the point that it was written. The past author at this point on the timeline has no memory of the book – it doesn't exist yet. Once the past author receives the book, he is experiencing something different, as a result of which the memory of the present author would have to change to be something different from its original understanding. In effect, the future that the present author occupies has become a different one after the experiment has taken place.

> **Einstein's theories are where you begin to run into fun paradoxes.**
>
> ELIZABETH HOWELL
> (2020)

This paradox is taken to a shocking extreme in Robert Heinlein's classic science fiction short story, *All You Zombies*, where someone becomes their own father and mother. This is the tightest of all bootstrap paradoxes, as the person who goes into the past constructs a tight loop in time. The central character is born with both male and female sex organs. Originally assigned as female, she is seduced by an older man. After giving birth to a baby, she is reassigned to be male due to medical complications. When the male version of the character is older, he tells a bartender about this. The bartender has a time machine and takes the male version back in time where the main character impregnates himself.

Nine months later, the bartender snatches the baby and takes it back in time to become the mother … and eventually, the bartender also proves to be an older version of the same character. With every character in the story shown to be the same individual, the bartender explains the story's title with a remark to his readers: 'I know where I came from – but where did all you zombies come from?' (Confusingly, Heinlein also wrote a novella called *By His Bootstraps*, but it is nowhere near as tightly crafted.)

Something from nothing

An argument that is sometimes used against backwards time travel comes from the first law of thermodynamics. This law says that in a closed system (we'll come back to that 'closed system'), mass-energy (because matter and energy can be converted one into the other) is conserved. In essence, to use an acronym that Robert Heinlein was fond of, TANSTAAFL: 'There ain't no such thing as a free lunch.' You can't have matter suddenly appearing; it has to have come from somewhere in some form.

It's certainly true that you can't make stuff appear from nowhere – but in practice this is a red herring. We know that objects can move in space and time. It happens all around us, without violating the first law. Seen from the point of view of the stuff, there is no violation and any other concern is really confusion due to the way that simultaneity is messed around by relativity. As a result of relativistic effects, while you can apparently get rich through time travel, in practice you will always fail – at least in terms of directly accumulating wealth.

> **The law of conservation ... rolls in music through the ages, and all terrestrial energy ... are [sic] but the modulations of its rhythm.**
>
> JOHN TYNDALL (1863)

Let's keep it simple and imagine I've got a 1 kg (2.2 lb) bar of platinum. At the time of writing that is worth around £21,000 ($27,000). I take it back to a time that's in my past, but that is after I first acquired the bar. Now, at that point in the past, I have two bars of platinum. I could wait a while, then take the two bars back and get four bars (and so on). But to avoid getting our brains into too much of a twist, let's stick to the original bar and one duplicate.

I'm now twice as rich. So I spend one of the bars and keep the other. I've somehow spent money that I got from nowhere. But there's a catch. Eventually I will reach the point in time when I took the bar back. If I don't take it back, then it will never have been duplicated. So I've got to take my one remaining platinum bar back in time. (If I spent both of the bars, I'd have to then get hold of another one to take back, or it would never have happened.) From the point of taking the bar back onwards, the future me has no platinum at all. Yes, I was able to spend a bar, but I had one bar all along. And I end up with no bars, having spent it.

This doesn't stop me making money through time travel, of course. I could invest the bar that I spend and make a profit in the future – but then I could have invested the original bar anyway. The only money-making ruse is the classic one that involves nothing more than information, so doesn't trouble the first law of thermodynamics. For example, I could send back the numbers of a winning lottery ticket to my earlier self before the draw was made. In practice, unless time machines could be kept secret, this would be a relatively short advantage, as was demonstrated for real with an earlier kind of time machine: the electric telegraph.

Before the telegraph was introduced it was common for bookmakers to take bets on races that had already finished because it could take hours or days for the results to travel from the racecourse to a distant city. When the telegraph made it possible to send results quickly, some gamblers realized this was the case before the bookies did and were able to clean up by betting on a result they already knew. Similarly, it's possible that lotteries and betting on the outcome of events would continue briefly once time travel was invented, allowing for a few to make a huge profit – but before long, the current modes of gambling would cease to exist.

It is difficult to envisage any traditional form of gambling, from lotteries to horse racing, surviving, as they all depend on the gambler predicting an outcome that is later announced to the world – so the result can be taken back to an earlier time to cheat the system. There would seem to be only one way in which such activities could continue. You could imagine a lottery where players are allocated random numbers but not told what those numbers are. That way, the owner of the random number could be contacted and paid their winnings without any means for an onlooker to relay the winning number back in time.

It's all in the entropy

As with the first law, the *second* law of thermodynamics is often invoked as a time-travel killer. The second law is far more interesting than the first law. As we saw on page 28, the second law effectively brings into existence a fixed direction in time. Although it was originally primarily about the movement of heat (hence 'thermodynamics' in its title), it is usually framed mathematically in terms of entropy.

UNDERSTANDING ENTROPY

Entropy is a measure of the disorder in a system. (A system here is anything being considered, from a pair of atoms up to a whole universe.) This sounds fuzzy, but has a precise definition. Mathematically, the entropy is discovered by finding the number of ways the components of the system can be rearranged. Specifically, the entropy S is calculated as

$$S = k \ln\Omega$$

Here k is Boltzmann's constant, ln signifies a natural log (logarithm to base e) and Ω is the number of possible configurations of the parts of the system.

The more ways those parts can be arranged, the higher the entropy. The second law of thermodynamics says that in a closed system – one where energy can't get in or out – the entropy will stay the same or increase.

The time-travel argument is that if you send back something containing information, like the book mentioned in the bootstrap paradox, then you are effectively decreasing entropy because information has lower entropy than a chaotic collection of components. However, the work-around here is that 'closed system' requirement in the statement of the law. By writing a book, for example, I do indeed manage to reduce entropy – but only at the cost of expending a lot of energy. Similarly, it would take a lot of energy to get the book (or whatever) back in time – so the second law argument is unlikely to present a problem.

Consistent histories

One possibility for the resolution of paradoxes is that this is simply how things are – get used to it. But there are two other possibilities. As we have already seen, so dramatic are the paradoxes of time travel that Stephen Hawking suggested his 'time COP' hypothesis that nature would act to prevent any paradoxical twist in time. The requirement, sometimes called 'consistent histories', initially sounds unlikely, as it seems to need the universe to consciously interfere in what we do. So, for example, if I planned to go back and save myself the effort of writing this book by giving a copy to an earlier self, we would expect the universe to say, 'Aha, that's not going to happen!' and prevent me.

However, if we take the consciousness out of it, this concept becomes less difficult. After all, we are quite happy that the universe tells matter what to do under the influence of gravity – why not also under the influence of time travel? In such a picture, things simply wouldn't happen if there was a possibility of generating a paradox.

This could occur in a number of ways. The time machine could fail, the time traveller could bounce back from just before creating the paradox, or, simplest of all, the time traveller would return from apparently setting up a paradox only to discover that nothing has changed – because the future was already locked in.

A different world

Alternatively, a theory from quantum physics could make such paradoxes perfectly acceptable.

We are familiar from science fiction with the ideas of parallel universes or alternative histories. But the concept doesn't entirely reside in the world of fiction. The many worlds hypothesis, which is an interpretation of quantum physics, suggests that every time a quantum particle has two possible choices, each choice occurs in a separate universe. If this were the case, the paradoxes collapse as each of the two conflicting parts takes place in a different universe.

If the many worlds hypothesis is true, then we have a get-out clause for the time traveller. Inevitably, the actions the time traveller takes in the past will result in different universes splitting off. If we take the example

> **The idea here is that when time travellers go back into the past, they enter alternative histories which differ from recorded history.**
>
> STEPHEN HAWKING (1988)

THE PROBLEM WITH QUANTUM PHYSICS

Quantum physics is a remarkable theory. It is astonishingly accurate in its predictions and it is the basis for all modern electronic technology. Yet it also portrays a strange picture of the world where quantum particles are nothing like the familiar objects we experience.

Quantum theory has a number of 'interpretations', intended to explain what is 'really' happening. The most frequently accepted is the Copenhagen interpretation, where particles exist only as probabilities until they interact with their surroundings and those probabilities coalesce into what is observed. This has now been developed to fix a problem known as waveform collapse, which meant it couldn't explain how particles settled on an actual position.

However, there are other interpretations, the most dramatic of which is the many worlds theory. Here, the universe effectively splits in two every time there could be different outcomes for a quantum particle, so all possible outcomes exist somewhere in one of a vast number of parallel universes.

of the grandfather paradox, the universe our traveller is born in, where her grandfather remained alive long enough to produce a family, is a different one from the world where the grandfather is killed. What isn't totally clear, though (leaving some scope for science fiction writers), is whether the traveller will return to the future where she doesn't exist or the future where the grandfather survived. Equally, she might return to a future where a different person was her grandfather to the one she killed because the original one is dead. If that were the case, she may not even remember who her 'original' grandfather was.

Although the many worlds hypothesis does provide this escape route from causal disaster, it ought to be stressed that many physicists do not accept this

hypothesis, which seems to make the universe unnecessarily complex. However, there is nothing as yet that can be used to disprove the hypothesis (its physical predictions are identical with those of other interpretations of quantum theory), so we can safely make use of it as our get-out clause.

Whatever the reality, the paradoxes of time are mind-boggling and surprisingly infrequently examined in film and TV science fiction. They make time travel both entertaining and a fascinating intellectual challenge. They would not arise with forwards time travel, something that we know is possible – it is just a matter of time before it can be scaled up, should it be desirable. Backwards time travel may never be practical, but making it happen is more an engineering problem than one of physical restraint.

Congratulations on completing the final lesson. You are now equipped to be a time traveller. Go out and enjoy the spacetime continuum.

GLOSSARY

Alcubierre drive – hypothetical spaceship drive that works by warping spacetime.

Antimatter – type of matter where the component particles have opposite values for a number of physical properties such as electrical charge.

Arrow of time – the idea that time has a preferred direction, distinguishing past and future, which seems dependent on the second law of thermodynamics.

Atomic clock – a very accurate clock that measures the passage of time from the decay of atomic nuclei.

Black hole – a star that has collapsed under such pressure that it effectively becomes a dimensionless point. Coming close enough to the black hole would make it impossible even for light to escape.

Bootstrap paradox – time-travel paradox where an object, information or a person is apparently created from nowhere.

Butterfly effect – concept from chaos theory based on the idea that a very small change in initial conditions (such as a butterfly stamping) can make a significant difference in the way events change through time.

Casimir effect – quantum effect where two very close flat plates feel a force pulling them together, in effect negative energy.

Causality – the idea of cause and effect, that one event can influence an outcome in the future, but not in the past.

Chaos – behaviour of a chaotic system, one where small differences in starting conditions result in major changes after the passage of time. The weather is a typical chaotic system.

Chronology projection conjecture – suggestion by English physicist Stephen Hawking that reality would conspire to prevent changes in the past that contradict the present.

Closed timelike loop – a warp in spacetime that produces a loop theoretically allowing backwards time travel.

Correlation – when two events appear to be linked, as one is close to the other spatially or in time, but it may or may not be true that one event caused the other.

Cosmic rays – high-energy particles that enter the Earth's atmosphere from outer space.

Cosmic strings – hypothetical, extremely dense structures in space that may have formed when the forces of nature split from each other at the beginning of the universe.

Cryonics – the controversial concept of storing human bodies or heads at extremely low temperatures immediately after death in the hope of reviving them in the future.

Dimension – a measurement scale in a particular direction of space or time. Dimensions are usually taken at right angles to each other, requiring three to cover all possible positions in space.

Dystopian futures – predictions of the future (or in science fiction) based on unpleasant consequences, coined as an opposite to a utopia.

Einstein–Rosen bridge – *see* Wormhole.

Electrochemical – a system that combines electricity and chemistry, such as the use of charged particles to carry signals in the brain.

Electromagnetic wave – a wave produced when a varying magnetic field produces a varying electrical field, which in turn produces another varying electrical field and so on. Light is an electromagnetic wave, though its wave-like properties are a quantum effect, rather than a traditional physical wave.

Electron – small fundamental particle, one or more of which is present on the outside of an atom, and which are the carriers of electrical current.

Encryption – the concealment of information by the use of codes or ciphers.

Entropy – a measure of the disorder in a system, determined by the number of ways its component parts can be organized. According to the second law of thermodynamics, in a closed system entropy will statistically stay the same or increase.

Equivalence principle – the idea that acceleration and the effects of gravity are indistinguishable, which led Einstein to develop his general theory of relativity.

Frame dragging – concept from the general theory of relativity that says that a moving massive body will tend to distort time and space in the same direction, so a rotating massive body can swirl spacetime around it.

Galilean relativity – basic theory of relativity developed by Galileo that shows that we need to consider the movement of the observer relative to the situation observed and that a steadily moving observer will see the same laws of physics as a stationary observer.

General theory of relativity – development of relativity by Albert Einstein that takes into account acceleration and explains gravity as the effect of a warping of space and time by matter. Shows that being near a massive object slows down time.

Global Positioning System (GPS) – system involving a network of satellites that broadcast time signals to enable locations on the surface of the Earth to be pinpointed.

Grandfather paradox – the outcome of travelling back in time to kill a grandparent before your parent was born. If it happened, would you still exist? And if not, it wouldn't have happened.

Hard science fiction – science fiction that as much as possible bases its speculative writing on science and technology that does not break the known physical laws.

International Space Station (ISS) – the longest-lasting space station to date, a joint US/Russian project, comprising a set of linked modules in low Earth orbit.

Ion – an atom that has gained or lost electrons to become electrically charged.

Kinetic energy – the energy due to the motion of an object.

Light clock – a clock where the 'tick' is a beam of light bouncing between a pair of parallel mirrors.

Many worlds hypothesis – an interpretation of quantum physics, where every time a quantum particle has two possible states the universe effectively splits into two, one for each state.

Muons – short-lived particles produced by high-energy collisions that are like electrons but with significantly more mass.

Neuron – cell in the brain or nervous system that can be connected to many others and processes electro-chemical signals.

Neutron star – a star that has collapsed to a compact, extremely dense structure only containing neutrons (the electrically neutral particles in atomic nuclei).

Newton's laws of motion – three basic laws, developed by Isaac Newton in the seventeenth century, which describe how forces acting on objects cause changes in their motion.

Nuclear fusion – the power source of the sun in which energy is generated when two or more lighter ions join together to make a heavier one, releasing energy in the process.

Paradox – the result of an apparently logical process that appears to be contradictory. A simple paradox is the statement 'This statement is untrue.' Is the statement true, or not?

Photoelectric effect – phenomenon where some light falling on metals or semiconductors can generate an electrical current.

Photon – quantum particle of light, and the carrier of the electromagnetic force. We might have been taught at school that light is a wave, but in fact it's a stream of photons with wave-like properties.

Prism – shape made by extending a triangle into the third dimension. Prisms are often made of glass or other transparent materials to experiment with reflection and refraction of light.

Quantum entanglement – mysterious quantum phenomenon where two particles can be separated to any distance and a change in one is immediately reflected in the other.

Quantum physics – the science of very small items such as electrons, atoms and photons, the behaviour of which is dramatically different to that of familiar objects, as many properties of these items, such as position, are not fixed definitively but exist as a range of probabilities until they interact with another object. Arises when aspects of nature are not continuously variable, but can only have values that vary by fixed amounts.

Quantum spin – a property of quantum particles that has some similarity to effects due to spinning, but in reality has nothing to do with rotation. When the spin of a particle is measured, it can only ever be either up or down.

Quantum tunnelling – the ability of quantum particles to pass through a barrier that should stop them, without spending time in that barrier.

Reaction mass – the material pushed out of the back of a rocket (or jet engine) that produces a forward thrust due to Newton's third law of motion.

Refraction – light phenomenon where a ray of light's direction of travel changes as it moves from one substance (e.g. air) to another (e.g. glass).

Relativity – the observation in physics that what is observed depends on the relative situation (in terms of space, time, movement and acceleration) of the observer.

Science fiction – sometimes called speculative fiction. Fiction that uses the vehicles of developments in science (which may or may not already exist) to explore human reaction to the worlds thus created.

Second law of thermodynamics – see Entropy.

Simultaneity – the idea that two events occur at the same point in time. According to special relativity, simultaneity is not absolute but relative, dependent on the relative motion of the observer and the events.

Spacetime – the combination of space and time that follows from the special theory of relativity's demonstration that the two are inherently entwined.

Special theory of relativity – development of Galilean relativity by Albert Einstein that takes into account that light always travels at the same speed, meaning that an observer will find that time on a moving object slows down and that the object increases in mass and shortens in the direction of movement.

Superluminal transmission – sending a signal faster than the speed of light, made possible by quantum tunnelling.

Tensor – a mathematical tool that maps one set of objects onto another, allowing a number of differential equations to be collapsed into a single equation.

Time dilation – the slowing of time due to relativistic effects that can enable travel through time.

Tipler cylinder – massive hypothetical cylinder constructed from neutron stars, which when rotated could enable backwards time travel due to frame dragging.

Twins paradox – arises when one twin is sent off into space at high speed for some time while the other remains on Earth. Because of time dilation, the space-travelling twin ends up younger than the earthbound twin.

Uncertainty principle – outcome of quantum theory that requires pairs of properties – for example position and momentum or energy and time – to be linked, so that the more accurately one item in a pair is known, the less accurately the other can be.

Uploading – the hypothetical process of scanning a human brain pattern and reproducing it in a computer in a way that it could be brought to conscious life in the future.

Virtual particles – because of the uncertainty principle, the energy in a location can vary so much in a short period of time that the energy can be sufficient to produce a pair of particles, one matter, one antimatter, which rapidly recombine and return to energy.

Vitrification – literally 'turning to glass', freezing a liquid as an amorphous solid, like glass, rather than as crystals, which are likely to damage any structures containing the liquid.

White hole – hypothetical anti-black hole, not unlike the state of the universe at the Big Bang.

Wormhole – hypothetical rip in spacetime linking one location to another without passing through the space in between.

FURTHER READING

For a deeper exploration of time travel:

Build Your Own Time Machine / How to Build a Time Machine by Brian Clegg, Duckworth (2011)/St Martin's Press (2011)

For a technical exploration of wormholes and other general relativity-related time travel:

Black Holes and Time Warps: Einstein's Outrageous Legacy by Kip Thorne, W. W. Norton (1994)

For a classic exploration of black holes and time:

A Brief History of Time: From the Big Bang to Black Holes by Stephen Hawking, Penguin Random House (1988)

For a better understanding of relativity:

The Reality Frame: Relativity and Our Place in the Universe by Brian Clegg, Icon Books (2017)

For an understanding of how time travel began in science fiction:

The Time Machine by H. G. Wells, Penguin Random House (1895) (published by William Heinemann in the US)

For the best picture of the place of time in modern physics:
Time Reborn: From the Crisis in Physics to the Future of the Universe by Lee Smolin, Allen Lane (2013)

For the best graphic exploration of time:
Introducing Time: A Graphic Guide by Craig Callender and Ralph Edney, Icon Books (2010)

For a detailed autobiography of Ronald Mallett:
The Time Traveller: One Man's Mission to Make Time Travel a Reality by Ronald Mallett and Brian Henderson, Doubleday (2007)

INDEX